Adsorption on Mesoporous Metal-Organic Frameworks in Solution for Clean Energy, Environment, and Healthcare

Adsorption on Mesoporous Metal-Organic Frameworks in Solution for Clean Energy, Environment, and Healthcare

Dr. Alexander Samokhvalov
Rutgers University, Camden, New Jersey, USA

CRC Press
Taylor & Francis Group
Boca Raton London New York

CRC Press is an imprint of the
Taylor & Francis Group, an **informa** business

CRC Press
Taylor & Francis Group
6000 Broken Sound Parkway NW, Suite 300
Boca Raton, FL 33487-2742

First issued in paperback 2019

© 2017 by Taylor and Francis Group, LLC
CRC Press is an imprint of Taylor & Francis Group, an Informa Business

No claim to original U.S. Government works

ISBN-13: 978-1-4987-6526-8 (hbk)
ISBN-13: 978-0-367-87787-3 (pbk)

Visit the Taylor & Francis Web site at
http://www.taylorandfrancis.com

and the CRC Press Web site at
http://www.crcpress.com

I would like to thank my beautiful wife, Olga, who was beside me throughout my career and when I was writing this book. Olga has been the source of my inspiration and motivation for continuing to improve my scientific knowledge. I dedicate this book to her.

I thank my daughter Masha for making me smile and for her kind understanding of those weekends and evenings when her dad was busy writing this book instead of playing games with her. I would also like to thank my parents and grandparents for allowing and encouraging me throughout my childhood to follow my ambitions to learn chemistry.

In the professional world, I sincerely thank Professor Ralph Yang from the Department of Chemical Engineering at the University of Michigan for believing in my academic potential and knowledge, which resulted in this book. I very much thank my past postdoctoral advisor, Professor Bruce Tatarchuk from the Chemical Engineering Department at Auburn University in Alabama, for demonstrating to me the profound and versatile connections between fundamental chemical science and the capabilities of chemical engineering that delivers goods to people. I am deeply grateful to my PhD advisor Professor Ron Naaman from the Department of Chemical Physics at the Weizmann Institute of Science in Israel for his great enthusiasm, creativity, and happiness in the systematic exploration of the unknown forces of nature and for encouraging my efforts in the same.

Finally, I thank everyone I have met in my career who provided their encouragement and help.

Contents

Preface

Adsorption and desorption in solution play significant roles in separation, detoxification of waste streams, purification, chromatography, heterogeneous catalysis, metabolism of medicinal drugs, and more. Metal–organic frameworks (MOFs) are well-ordered 3D hybrid organic–inorganic polymers that contain metal cations and structure-building organic "linker" units. Mesoporous MOFs with pore sizes 2–50 nm are particularly suitable for adsorption and adsorption-based separations of large molecules of organic and bio-organic compounds.

Thousands of organic compounds and in particular aromatic and heterocyclic compounds are widely used as feedstock for industrial chemical synthesis and as fine chemicals, major components of liquid fossil fuels, dyestuffs, industrial solvents, agricultural chemicals, medicinal drugs, pharmaceuticals and personal care products (PPCPs), active pharmaceutical ingredients (APIs), and so on.

There is strong interest in the synthesis, characterization, and studies of both known and newly synthesized mesoporous MOFs for adsorption in solution to achieve high adsorption capacity, selectivity, and the possibility of multiple regeneration of the "spent" sorbent.

This book is intended to cover experimental fundamental research on using mesoporous MOFs for emerging applications of major industrial, environmental, and academic importance, especially for purification and separations of water and liquid fossil fuels.

List of Figures

List of Tables

Author

Alexander Samokhvalov received his BSc and MSc in chemistry at the Novosibirsk State University in Russia. He earned his PhD in chemistry at the Weizmann Institute of Science in Israel. He had spent a few years of postdoctoral training in the United States at Duke University, UC Santa Barbara, and Auburn University in Alabama. Since 2010, he is an assistant professor of physical chemistry at the Chemistry Department of Rutgers University. His research interests are in mechanistic studies of adsorption by metal–organic frameworks (MOFs) in solution.

1 Introduction

Metal–organic frameworks (MOFs) are well-ordered 3D hybrid organic–inorganic polymers that contain metal cations and structure-building organic "linker" units. The generic structure of the MOF can be seen in Figure 1.1 (Janiak and Vieth 2010).

MOFs are of particular interest to chemists, physicists, and chemical engineers. They exhibit a unique combination of favorable structural and chemical parameters, including (1) a well-defined nanostructure that is created and controlled by chemical synthesis; (2) a very high total surface area up to some 5000 m^2/g; (3) a rather small "dead volume" not accessible by potential adsorbate molecules; (4) the presence of metal cations that can serve as active sites for adsorption and/or catalytic reactions, and (5) a moderate-to-high stability to ambient factors such as temperature, oxygen, atmospheric humidity, and solvents including water. In the recent decade, a rather strong research interest has been directed toward the synthesis of new MOFs, the determination of their chemical composition and structure, and investigations of their potential applications in adsorption (Khan et al. 2013), separations (Li et al. 2012), chromatographic analysis (Chang et al. 2012, Xie and Yuan 2013, Yu et al. 2013), heterogeneous catalysis (Corma et al. 2010, Dhakshinamoorthy et al. 2011, 2014), and their biomedical applications as drug carriers (Huxford et al. 2010, Keskin and Kizilel 2011, Horcajada et al. 2012). The interest of the chemical industry toward commercialization of technologies relying on MOFs has been steadily increasing (Mueller et al. 2006, Czaja et al. 2009).

MOFs are classified as microporous, mesoporous, and macroporous. According to the standard definition by the International Union of Pure and Applied Chemistry (IUPAC), mesoporous MOFs have pore sizes in the range 2–50 nm (Everett 1972). The sizes of mesocavities in the mesoporous MOFs are usually comparable to or larger than the molecular sizes of the "typical" molecules of organic compounds, except polymers. Therefore, mesoporous MOFs are particularly promising for adsorption of many organic compounds of major industrial, environmental, health-care, and academic interest.

Figure 1.2 shows the increase in the number of research papers published annually on mesoporous MOFs since 2002, as was found in the Web of Science (WOS) citation database by Thomson Reuters, retrieved on April 16, 2016. The search topic was "mesoporous AND metal-organic framework," and 1040 records were found.

Still, much less work has been devoted to mesoporous MOFs, compared to microporous MOFs. Few review papers have been published on the synthesis and structural aspects of mesoporous MOFs. The review paper of 2010 (Fang et al. 2010) discusses the research on mesoporous MOFs, with a focus on structural aspects: 3D channels, 1D channels, large cavities, and MOFs based on supramolecular templates. The review paper of 2012 (Song et al. 2012) discusses the more recent developments in the design of mesoporous MOFs and certain applications, such as gas storage, catalysis, adsorption of volatile organic compounds (VOCs), and drug delivery. Another review

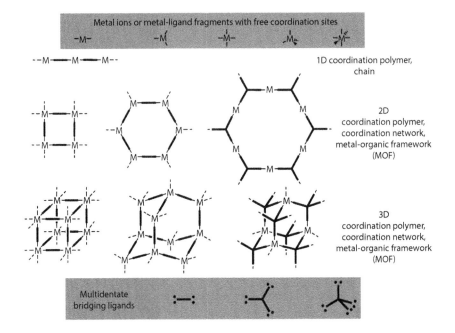

FIGURE 1.1 Schematic presentation for the construction of typical coordination polymers/ MOFs from molecular building blocks. (From Janiak, C. and Vieth, J.K., MOFs, MILs and more: Concepts, properties, and applications for porous coordination networks (PCNs), *New J. Chem.*, 34(11), 2366–2388, 2010. Reproduced by permission of the Centre National de la Recherche Scientifique (CNRS) and The Royal Society of Chemistry.)

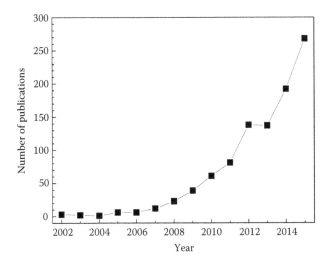

FIGURE 1.2 The number of research papers published annually on mesoporous MOFs.

of 2012 (Xuan et al. 2012) covers advances in the studies of mesoporous MOFs: various types based on their synthesis, activation of porosity, surface modification, and potential applications in storage and separation in the gas phase, in catalysis, in drug delivery, and in imaging. We published the review paper on adsorption of aromatic and heterocyclic compounds on mesoporous MOFs in solution (Samokhvalov 2015). In recent years, the particular importance of mesoporous MOFs has been widely recognized by the international research community as can be seen by the review papers published in languages other than English (e.g., Song et al. 2014).

MOFs including mesoporous MOFs have a moderate-to-high stability to elevated temperatures and oxygen, while their hydrolysis represents a common problem (e.g., Bezverkhyy et al. 2014). Possibly, a well-known hydrolysis of MOFs explains why adsorption on the MOFs in the liquid phase has been studied much less, compared to the adsorption in the gas phase.

This book provides a systematic and critical analysis relating to chemical and structural aspects of adsorption of various chemical compounds on mesoporous MOFs in solution. The effects of postsynthetic modifications (PSMs) of mesoporous MOFs on their adsorption capacity, selectivity, and the mechanisms of adsorption are analyzed. Adsorption in aqueous solutions is considered of major importance for emerging industrial-scale applications. Therefore, adsorption and desorption of individual chemical compounds of interest and their commercially useful mixtures (e.g., liquid fossil fuels) have been discussed in this book as separate chapters for aqueous and nonaqueous solutions. The stability of mesoporous MOFs in liquid water and in water vapor is critically discussed, and mechanisms of (in)stability are analyzed.

Adsorption and desorption of chemical compounds of the following classes are discussed: organic dyes, medicinal drugs and active pharmaceutical ingredients (APIs), aromatic sulfur compounds, aromatic nitrogen heterocyclic compounds from petroleum and liquid fossil fuels, organic compounds as feedstock for industrial chemical synthesis, aromatic hydrocarbons, chlorinated aromatic compounds, and miscellaneous inorganic compounds. The following molecular-level mechanisms of adsorption and desorption are discussed in this book: bonding by coordination, Brønsted acid–base interactions, hydrogen bonding, Lewis acid–base interactions, ionic interactions, π–π interactions, and nonspecific interactions. The impetus has been placed on the spectroscopic characterization of adsorbate molecules and chemical bonds formed with the adsorption sites, as spectroscopic studies are very useful in mechanistic research.

Due to the rapid progress in this "hot" research field in the recent decade, this book is unlikely to cover all publications on the topic. Instead, it aims to provide a systematic and critical overview of major achievements, trends, and challenges. To encourage further research in this quickly growing field, we attempted to identify hidden "niches" where, to our knowledge, application-oriented and/or mechanistic studies have not been published yet. Perspectives and limitations of a wide use of mesoporous MOFs as industrial-scale sorbents are also discussed. This book can serve as a reference handbook for readers in various subfields of chemistry, chemical engineering, materials science, biomedical science, environmental research, and for professionals in academe and industry who work with advanced porous functional materials.

REFERENCES

Bezverkhyy, I., G. Ortiz, G. Chaplais, C. Marichal, G. Weber, and J. P. Bellat. 2014. MIL-53(Al) under reflux in water: Formation of gamma-AlO(OH) shell and H_2BDC molecules intercalated into the pores. *Microporous and Mesoporous Materials* 183:156–161.

Chang, C. L., X. Wang, Y. Bai, and H. W. Liu. 2012. Applications of nanomaterials in enantioseparation and related techniques. *Trac-Trends in Analytical Chemistry* 39:195–206.

Corma, A., H. Garcia, and F. X. Llabres i Xamena. 2010. Engineering metal organic frameworks for heterogeneous catalysis. *Chemical Reviews* 110(8):4606–4655.

Czaja, A. U., N. Trukhan, and U. Mueller. 2009. Industrial applications of metal-organic frameworks. *Chemical Society Reviews* 38(5):1284–1293.

Dhakshinamoorthy, A., M. Alvaro, and H. Garcia. 2011. Metal-organic frameworks as heterogeneous catalysts for oxidation reactions. *Catalysis Science & Technology* 1(6):856–867.

Dhakshinamoorthy, A., A. M. Asiri, and H. Garcia. 2014. Catalysis by metal-organic frameworks in water. *Chemical Communications* 50(85):12800–12814.

Everett, D. H. 1972. Manual of symbols and terminology for physicochemical quantities and units, Appendix II: Definitions, terminology and symbols in colloid and surface chemistry. *Pure and Applied Chemistry* 31(4):577–638.

Fang, Q. R., T. A. Makal, M. D. Young, and H. C. Zhou. 2010. Recent advances in the study of mesoporous metal-organic frameworks. *Comments on Inorganic Chemistry* 31(5–6):165–195.

Horcajada, P., R. Gref, T. Baati, P. K. Allan, G. Maurin, P. Couvreur, G. Ferey, R. E. Morris, and C. Serre. 2012. Metal-organic frameworks in biomedicine. *Chemical Reviews* 112(2):1232–1268.

Huxford, R. C., J. D. Rocca, and W. Lin. 2010. Metal–organic frameworks as potential drug carriers. *Current Opinion in Chemical Biology* 14(2):262–268.

Janiak, C. and J. K. Vieth. 2010. MOFs, MILs and more: concepts, properties and applications for porous coordination networks (PCNs). *New Journal of Chemistry* 34(11):2366–2388.

Keskin, S. and S. Kizilel. 2011. Biomedical applications of metal organic frameworks. *Industrial & Engineering Chemistry Research* 50(4):1799–1812.

Khan, N. A., Z. Hasan, and S. H. Jhung. 2013. Adsorptive removal of hazardous materials using metal-organic frameworks (MOFs): A review. *Journal of Hazardous Materials* 244:444–456.

Li, J. R., J. Sculley, and H. C. Zhou. 2012. Metal-organic frameworks for separations. *Chemical Reviews* 112(2):869–932.

Mueller, U., M. Schubert, F. Teich, H. Puetter, K. Schierle-Arndt, and J. Pastre. 2006. Metal-organic frameworks-prospective industrial applications *Journal of Materials Chemistry* 16(7):626–636.

Samokhvalov, A. 2015. Adsorption on mesoporous metal–organic frameworks in solution: aromatic and heterocyclic compounds. *Chemistry—A European Journal* 21(47):16726–16742.

Song, L. F., H. Y. Xia, H. X. Chen, Z. Li, and J. J. Lu. 2014. Preparation and application of mesoporous metal-organic frameworks. *Progress in Chemistry* 26(7):1132–1142.

Song, L. F., J. Zhang, L. X. Sun, F. Xu, F. Li, H. Z. Zhang, X. L. Si et al. 2012. Mesoporous metal-organic frameworks: design and applications. *Energy & Environmental Science* 5(6):7508–7520.

Xie, S. M. and L. M. Yuan. 2013. Metal-organic frameworks used as chromatographic stationary phases. *Progress in Chemistry* 25(10):1763–1770.

Xuan, W., C. Zhu, Y. Liu, and Y. Cui. 2012. Mesoporous metal-organic framework materials. *Chemical Society Reviews* 41(5):1677–1695.

Yu, Y. B., Y. Q. Ren, W. Shen, H. M. Deng, and Z. Q. Gao. 2013. Applications of metal-organic frameworks as stationary phases in chromatography. *Trac-Trends in Analytical Chemistry* 50:33–41.

2 Postsynthetic Modifications of Mesoporous MOFs for Adsorption-Based Applications

The postsynthetic modifications (PSMs) of metal–organic frameworks (MOFs) have been periodically and quite extensively reviewed (e.g., Tanabe and Cohen 2011, Cohen 2012, Deria et al. 2014). One of the major advantages of PSMs of MOFs is the elimination of the possibility for the functional groups in the precursor of the linker to interfere with each other during synthesis. PSMs can be conducted by chemical modification of the linker unit and/or the metal-containing node of the structure of the target MOF. The following kinds of PSMs were reported to chemically transform the linker or the functional group in the linker: (1) covalent modification, (2) deprotection of the functional group in the linker, and (3) chemical reduction. The PSM of the metal node includes (1) incorporation of the nonframework aka "pendant ligands" via bonding to the coordinatively unsaturated metal sites (CUS), (2) grafting alkyl or silyl groups to the oxygen atoms in the metal oxide nodes of the MOF, and (3) attachment of metal cations or coordination compounds at the node oxygen sites via atomic layer deposition (ALD) or chemical reaction with organometallic precursors in solution. The currently available methods for PSMs are quite diverse, and chemical details can be found in multiple review papers. This chapter discusses those PSM methods that have found frequent application in the synthesis of mesoporous MOFs for different applications, specifically as sorbents and heterogeneous catalysts. It is convenient to start by reviewing the PSMs of the most common mesoporous MOFs.

2.1 PSMs OF MIL-101 FOR ADSORPTION AND CATALYSIS IN SOLUTION

Arguably, the most well-known mesoporous MOF is MIL-101(Cr). In MIL 101(Cr), Cr^{3+} cations are connected to the linker units, the anions of 1,4-benzenedicarboxylic acid (BDC, Figure 2.1, Hong et al. 2009). In Fig. 2.1, references [51, 52] are the two early papers on MIL-101(Cr), namely Férey et al. (2005a) and Férey et al. (2005b). The structure of MIL-101 contains two mesocages present in a 2:1 ratio. The smaller mesocages have an internal diameter of 29 Å and contain only pentagonal windows with a free aperture of ca. 12 Å. On the other hand, the large mesocages with an

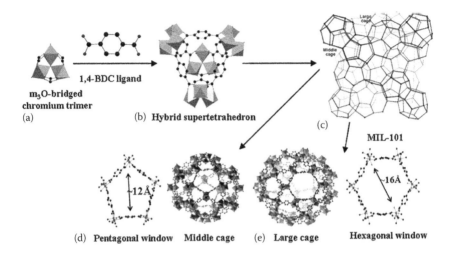

FIGURE 2.1 Basic building units and crystal structure of MIL-101 [51,52]: (a) Original building block with a trimer of chromium octahedral chelated by two carboxylic functions of organic ligand, (b) hybrid supertetrahedron formed by using terephthalic acid, which occupies the edges of the supertetrahedron, and (c) a polyhedral representation of the 3D organization in the zeotype architecture of MIL-101 comprising two cages, which are present in a 2:1 ratio of middle and large cages—(d) middle cages (20 ST, an internal diameter of ≈29 Å) with only pentagonal windows (free aperture ≈12 Å) and (e) large cages (28 ST, internal diameter: 34 Å) with 12 pentagonal and 4 hexagonal windows (free aperture ≈14.5 Å × 16 Å). (Reproduced from Hong, D.Y. et al., *Adv. Funct. Mater.*, 19(10), 1537, 2009. With permission.)

TABLE 2.1
Postsynthetic Modifications of MIL-101(Cr) Based on Chemical Reactions of Adsorption

PSM Agent	Modified MOF	Application	Reference
Pd(acac)$_2$	Pd/MIL-101(Cr)	Catalytic hydrogenation of styrene	Henschel et al. (2011)
Pd(NO$_3$)$_2$	Pd/MIL-101(Cr)	Catalytic C$_2$ arylation of indoles in solution	Huang et al. (2011)
H$_2$PdCl$_4$	Pd@MIL-101(Cr)(F)	Catalytic reduction of 4-nitrophenol	Yadav et al. (2012)
H$_2$PtCl$_6$	Pt@MIL-101(Cr)(F)	Catalytic liquid-phase ammonia borane hydrolysis	Aijaz et al. (2012)
Phosphotungstic acid, PWA	PWA/MIL-101(Cr)(F)	Adsorption of N-aromatics	Ahmed et al. (2013b)
Graphene oxide, GO	GO/MIL-101(Cr)(F)	Adsorption of N-aromatics	Ahmed et al. (2013c)
Acidic ionic liquid, IL	IL/MIL-101(Cr)	Adsorption of aromatic S compounds	Khan et al. (2014)

internal diameter of 34 Å contain 12 pentagonal and 4 hexagonal windows with a free aperture of 14.5 Å × 16 Å.

MIL-101(Cr) belongs to the family of isostructural MIL-101 MOFs, which also include MIL-101(Fe) (Čendak et al. 2014), MIL-101(Al) (Serra-Crespo et al. 2011), and MIL-101(Sc) (Mitchell et al. 2013). Table 2.1 shows published data on the most common PSMs of MIL-101(Cr) based upon chemisorption.

An activated MIL-101(Cr) was impregnated with a chloroform solution of palladium acetylacetonate Pd(acac)$_2$ via an incipient wetness impregnation technique, that is, by adsorption (Henschel et al. 2011). Then, the obtained precursor Pd(acac)$_2$/MIL-101(Cr) was heated in a flow of hydrogen at 473 K for 1 h to obtain the desired Pd/MIL-101(Cr) material. The obtained Pd/MIL-101(Cr) was studied for adsorption of styrene and ethyl cinnamate in solution, and for catalytic hydrogenation of styrene to ethylbenzene in the liquid phase.

The supported catalyst Pd/MIL-101(Cr) was prepared via infiltration of Pd(NO$_3$)$_2$ solution into the activated MIL-101(Cr), followed by hydrogen reduction at 200°C (Huang et al. 2011). The structure of MIL-101(Cr) did not change after loading up to 0.5 wt% Pd. Transmission electron microscopy (TEM) indicated that the majority of metallic Pd nanoparticles had a mean diameter of 2.6 ± 0.5 nm and they were enclosed in mesocavities of MIL-101(Cr). The obtained composite material Pd/MIL-101(Cr) has been successfully used in the catalytic C$_2$ arylation of indoles in dimethylformamide (DMF) at 120°C.

The highly monodispersed Pd nanoparticles have been incorporated inside the mesopores of MIL-101(Cr) by the novel "double solvents" method (Aijaz et al. 2012, Yadav et al. 2012). This method has been found particularly suitable for avoiding an aggregation of metal nanoparticles on the external surfaces of an MIL-101(Cr) support. Specifically, cyclohexane was used as a hydrophobic solvent to disperse the MOF, while an aqueous solution of PdCl$_2$ precursor was used as a "hydrophilic solvent" to enter and fill the mesopores. Apparently, selective adsorption on the hydrophilic interior of the mesopores was favored because of the high polarity of the solvent. The obtained Pd@MIL-101(Cr) has been found to be highly catalytically active in the reduction of 4-nitrophenol with sodium borohydride (Yadav et al. 2012).

A similar procedure has been used for the synthesis of Pt@MIL-101(Cr) (Figure 2.2).

MIL-101(Cr) was impregnated with phosphotungstic acid (PWA) to form a composite sorbent PWA/MIL-101(Cr)(F) with acidic surface groups (Ahmed et al. 2013b). In a similar fashion, MIL-101(Cr)(F) was treated with graphene oxide (GO) in order to form composite material GO/MIL-101(Cr)(F) for adsorptive denitrogenation (Ahmed et al. 2013c) and with ionic liquid (IL) to form IL/MIL-101(Cr) for adsorptive desulfurization of liquid fossil fuels (Khan et al. 2014).

The second major kind of PSM is covalent modification of the aromatic ring in the linker of MIL-101(Cr) (see Table 2.2).

Nitration of MIL-101(Cr)(F) with a mixture of nitric and sulfuric acids resulted in the introduction of the nitro group to the aromatic ring of the BDC linker (Bernt et al. 2011, Figure 2.3).

Furthermore, the nitro group in the BDC linker of the obtained MIL-101(Cr)(F)-NO$_2$ was reduced with SnCl$_2$ in ethanol to the amino group in the linker (Bernt et al. 2011).

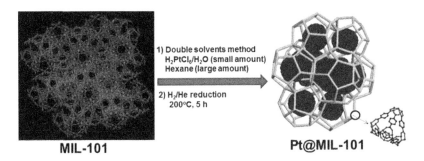

MIL-101 1) Double solvents method
H$_2$PtCl$_6$/H$_2$O (small amount)
Hexane (large amount)

2) H$_2$/He reduction
200°C, 5 h

Pt@MIL-101

FIGURE 2.2 Schematic representation of the synthesis of Pt nanoparticles inside the MIL-101 matrix using the "double solvents" method. (Reprinted with permission from Aijaz, A., Karkamkar, A., Choi, Y.J., Tsumori, N., Rönnebro, E., Autrey, T., Shioyama, H., and Xu, Q., Immobilizing highly catalytically active Pt nanoparticles inside the pores of metal–organic framework: A double solvents approach, *J. Am. Chem. Soc.*, 134(34), 13926–13929. Copyright 2012 American Chemical Society.)

TABLE 2.2
PSMs of MIL-101 via Covalent Bonding to the –NH$_2$ Group in the Linker

MOF	PSM Agent	Modified MOF	Reference
MIL-101(Cr)(F)-NH$_2$	Ethyl isocyanate	Figure 2.4	Bernt et al. (2011)
MIL-101(Cr)(F)–NH$_2$	*p*-Phenylazobenzoylchloride	Figure 2.5	Modrow et al. (2012)
MIL-101(Cr)(F)–NH$_2$	4-(Phenylazo)-phenylisocyanate	Figure 2.5	Modrow et al. (2012)
MIL-101(Al)–NH$_2$	4-Methylaminopyridine (4-MAP) in the combination with disuccinimidyl suberate (DSS)	Figure 2.6	Bromberg et al. (2012)
MIL-101(Al)–NH$_2$	Phenyl isocyanate	Figure 2.7	Wittmann et al. (2015)
MIL-101(Al)-NH$_2$	HO-Pro		Bonnefoy et al. (2015)
	HO-Gly-Pro		
	HO-Gly-Gly		
	HO-Gly-Gly-Gly	Figure 2.8	
	HO-Sar-Gly-Ala		
	HO-Gly-Gly-Gly-Ala		
	HO-Gly-Phe-Gly-Gly		

In order to introduce Brønsted acidity to MOF functional materials, sulfation of MIL-101(Cr)(F) was conducted using a mixture of trifluoromethanesulfonic (triflic) anhydride and sulfuric acid (Goesten et al. 2011). The obtained sulfated MIL-101(Cr)(F) was characterized by solid-state nuclear magnetic resonance (NMR) spectroscopy, X-ray absorption near-edge structure (XANES) spectroscopy, and Fourier transform infrared (FTIR) spectroscopy. The acidic sulfoxy group was found to be attached in up to 50% of the present BDC linkers. The obtained sulfated MIL-101(Cr) was found active in the acid-catalyzed esterification of n-butanol with acetic acid (Goesten et al. 2011). Later, similarly prepared sulfated MIL-101(Cr)(F) was

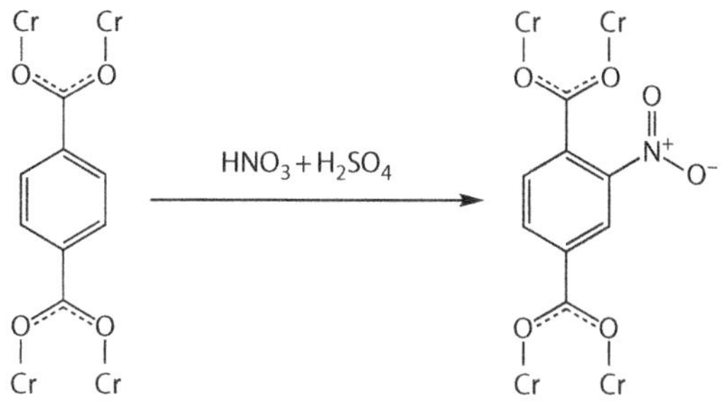

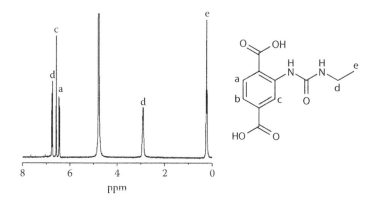

FIGURE 2.3 Schematic representation of the post-synthetic modification of Cr-MIL-101. (Adapted from Bernt, S., Guillerm, V., Serre, C., and Stock, N., Direct covalent post-synthetic chemical modification of Cr-MIL-101 using nitrating acid, *Chem. Commun.*, 47(10), 2838–2840, 2011. Reproduced by permission of The Royal Society of Chemistry.)

used for selective adsorption of basic nitrogen heterocyclic compounds from model liquid petroleum fuel (Wang et al. 2013).

The amino group in the BDC linker of MIL-101(Cr)(F)-NH$_2$ attached to the benzene ring via nitration and subsequent reduction as described earlier (Bernt et al. 2011) was found to be quite useful in further PSM with various additional functional groups connected via covalent bonds to this amino group. Specifically, the chemical reaction of MIL-101(Cr)(F)-NH$_2$ with ethyl isocyanate in acetonitrile resulted in the modified MIL-101 shown in Figure 2.4.

Azo functionality was introduced via covalent chemical bonds into MIL-101(Cr)(F)-NH$_2$ (Modrow et al. 2012). The MOF denoted Cr-MIL-101-NH$_2$ was further postsynthetically modified with p-phenylazobenzoylchloride (1) and 4-(phenylazo)-phenylisocyanate (2) (Figure 2.5).

FIGURE 2.4 ^{1}H NMR spectrum of the linker of Cr-MIL-101–UR$_2$ measured in NaOD/D$_2$O (20%). The signals can be clearly assigned and the integrals also fit. The signal at 4.8 ppm can be assigned to water. (From Bernt, S., Guillerm, V., Serre, C., and Stock, N., Direct covalent post-synthetic chemical modification of Cr-MIL-101 using nitrating acid, *Chem. Commun.*, 47(10), 2838–2840, 2011. Reproduced by permission of The Royal Society of Chemistry.)

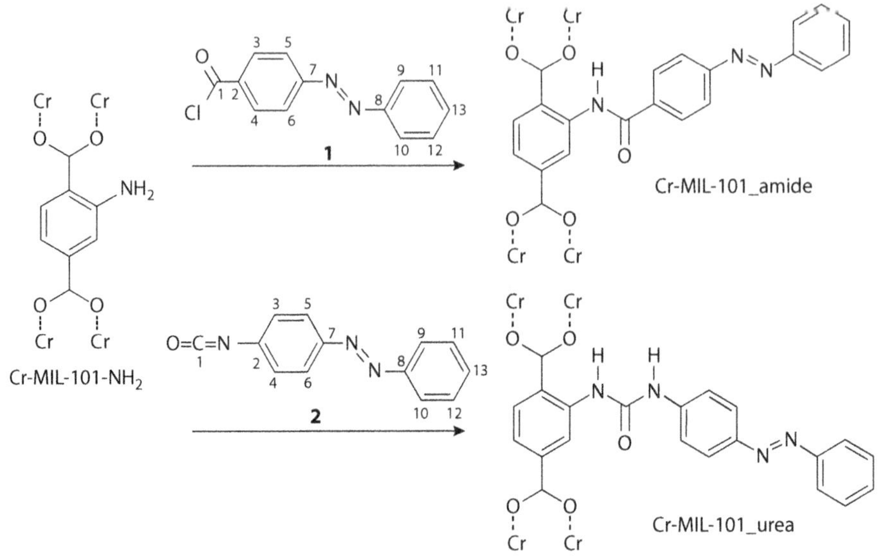

FIGURE 2.5 Post-synthetic modification of Cr-MIL-101-NH$_2$ with p-phenylazobenzo-ylchloride (**1**) and 4-(phenylazo)phenylisocyanate (**2**). (From Modrow, A., Zargarani, D., Herges, R., and Stock, N., Introducing a photo-switchable azo-functionality inside Cr-MIL-101-NH$_2$ by covalent post-synthetic modification, *Dalton Trans.*, 41(28), 8690–8696, 2012. Reproduced by permission of The Royal Society of Chemistry.)

The obtained functionalized MOFs denoted Cr-MIL-101_amide and Cr-MIL-101_urea have the azo functional groups inside the mesoporous cages of the MIL-101 structure. Powder x-ray diffraction (XRD) and nitrogen adsorption measurements confirmed the preservation of the framework of MIL-101, and successful chemical modification was confirmed by FTIR and NMR spectroscopy. Furthermore, the cis/trans isomerization of the azobenzene unit in Cr-MIL-101_amide and Cr-MIL-101_urea products of the PSM was achieved by irradiation with light, and the completion of the reaction was confirmed by UV/Vis absorption spectroscopy. The observed *cis/trans* isomerization resulted in different adsorption capacities for the two respective isomers, as has been revealed by the adsorption of methane (Modrow et al. 2012).

The PSM of the amino group in the BDC linker was extended to MIL-101 beyond the "benchmark" material MIL-101(Cr)(F). The amino groups in MIL-101(Al)-NH$_2$ were functionalized with covalent bonds using 4-methylaminopyridine (4-MAP) in combination with disuccinimidyl suberate (DSS) conjugation (Bromberg et al. 2012). Figure 2.6 shows the simplified reaction scheme, where 2-aminoterephthalic acid (2-ATA) mimics the respective linker in MIL-101(Al)-NH$_2$.

The thin films formed by the PSM-treated MOF shown in Figure 2.6 were studied in the catalytic hydrolysis of diisopropyl fluorophosphate (DFP), the simulant of the chemical warfare agent, in an aqueous solution (Bromberg et al. 2012).

The stability of MOFs toward water is a critical issue that significantly limits wide-scale industrial use of MOFs, including mesoporous MOFs. Recently, simple covalent PSM of mesoporous MIL-101(Al)-NH$_2$, also denoted as Al-MIL-101-NH$_2$,

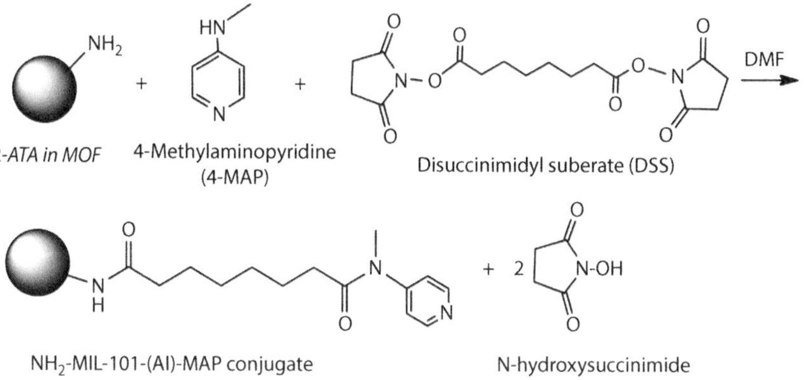

FIGURE 2.6 Reaction of 2-aminoterephthalic acid (2-ATA) and 4-methylaminopyridine (4-MAP) conjugation using DSS as an activating agent. (Reprinted with permission from Bromberg, L., Klichko, Y., Chang, E.P., Speakman, S., Straut, C.M., Wilusz, E., and Hatton, T.A., Alkylaminopyridine-modified aluminum aminoterephthalate metal-organic frameworks as components of reactive self-detoxifying materials, *ACS Appl. Mater. Interfaces*, 4(9), 4595–4602. Copyright 2012 American Chemical Society.)

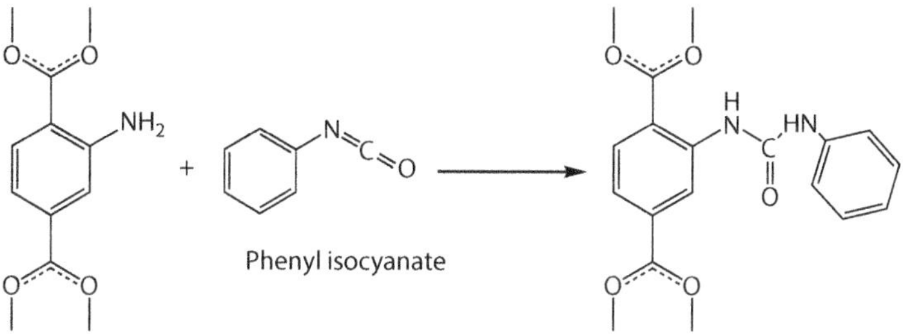

FIGURE 2.7 Post-synthetic modification of Al-MIL-101-NH$_2$ by treatment with phenyl-isocyanate. (Adapted from Wittmann, T., Siegel, R., Reimer, N., Milius, W., Stock, N., and Senker, J.: Enhancing the water stability of Al-MIL-101-NH$_2$ via postsynthetic modification. *Chem. A Eur. J.* 2015. 21(1). 314–323. Copyright Wiley-VCH Verlag GmbH & Co. KGaA. Reproduced with permission.)

was reported to result in Al-MIL-101-URPh by using phenyl isocyanate (Wittmann et al. 2015, Figure 2.7).

Upon PSM with phenyl isocyanate, about 86% of the amino groups present in MIL-101(Al)-NH$_2$ were converted into phenylurea structural units (Wittmann et al 2015). The modified MOF was characterized by chemical analysis, adsorption of argon, nitrogen and water vapors, powder XRD, thermogravimetric analysis (TGA), solid-state NMR, and FTIR spectroscopy. The PSM resulted in a decrease in the Brunauer–Emmett–Teller (BET) total surface area from 3363 to 1555 m^2/g, but the MIL-101 structure was mostly preserved. Since the –NH$_2$ groups were converted to the nonbasic amide groups, the interactions with water molecules were found to

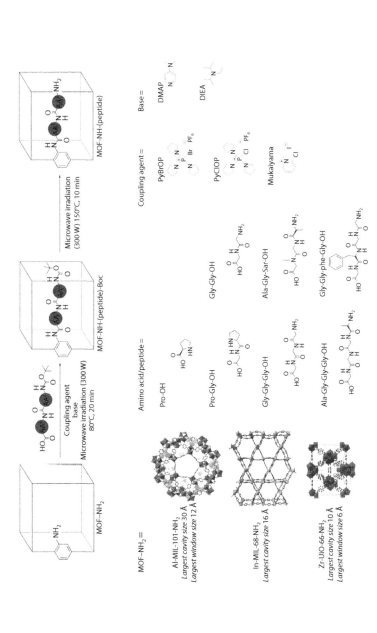

FIGURE 2.8 Parameters investigated for the optimization of the two-step peptide grafting process into various MOFs.[a] [a]*Note:* Reverse nomenclature is used for isolated peptides and MOF-grafted peptides: for example, Pro-Gly-OH, in which the amino acid–bearing terminal NH is the first listed, is grafted to give Al-MIL-101-NH-Gly-Pro, in which the amino acid–bearing terminal NH becomes the last one listed. Pro = proline, Gly = glycine, Sar = sarcosine, Ala = alanine, and Phe = phenylalanine. (Reprinted with permission from Bonnefoy, J., Legrand, A., Quadrelli, E.A., Canivet, J., and Farrusseng, D., Enantiopure peptide-functionalized metal–organic frameworks. *J. Am. Chem. Soc.,* 137(29), 9409–9416. Copyright 2015 American Chemical Society.)

be much weaker than in MIL-101(Al)-NH$_2$. The obtained Al-MIL-101-URPh was found to be stable in liquid water for over one week (Wittmann et al. 2015).

In a recent paper, the PSM of the amino group in several MOFs, including Al-MIL-101-NH$_2$, was reported using peptides (Bonnefoy et al. 2015), Figure 2.8.

The reported method used microwave-assisted PSM and resulted in enantiopure peptides anchored inside the cavities of the respective MOFs. For Al-MIL-101-NH$_2$, the yield varied in the range 5%–60%, depending on the peptide used (Figure 2.8).

Diazonium salts RN$_2$$^+X^-$ where R is an aryl group are synthesized by the reaction of primary aromatic amines with sodium nitrite and acid in the aqueous solution. Diazonium salts can further react with a range of anions, including halogen anions, thus forming the respective ring-substituted aromatic compounds. This strategy has been applied to convert the amino group in MIL-101-NH$_2$ to the diazo group MIL-101-N$_2$$^+$ (Jiang et al. 2012). The obtained MIL-101-N$_2$$^+$ was then converted to the MOFs with the halogen or the azo group containing substituents in the aromatic ring of the BDC linker (Figure 2.9).

This method is termed "tandem" postsynthetic modification. Specifically, in synthetic step #1, an arene diazonium chloride salt was prepared in situ by suspending MIL-101(Cr)-NH$_2$ in the aqueous solution of an acid, followed by addition of NaNO$_2$ and stirring at 0°C–5°C. Next, the required source of the anion was added; for instance, iodine-functionalized MIL-101(Cr)-I was prepared by adding the aqueous solution of KI to the reaction mixture and stirring. MIL-101(Cr)-azo was prepared by adding an aqueous solution of phenol to the diazonium chloride of MIL-101, followed by adjustment of the pH 7.0 and stirring at room temperature (Jiang et al. 2012).

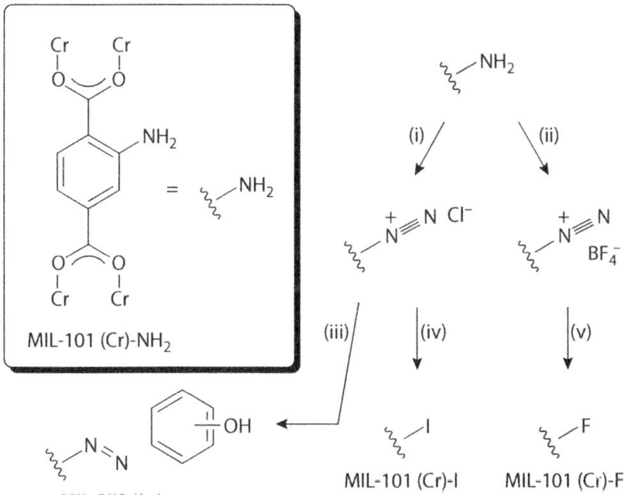

FIGURE 2.9 Tandem post-synthetic modification of MIL-101(Cr)-NH$_2$. (i) NaNO$_2$, HCl (aq); (ii) NaNO$_2$, HBF$_4$ (aq); (iii) C$_6$H$_5$OH (aq), Na$_2$CO$_3$; (iv) KI (aq); (v) 100°C. (From Jiang, D., Keenan, L.L., Burrows, A.D., and Edler, K.J., Synthesis and post-synthetic modification of MIL-101(Cr)-NH$_2$ via a tandem diazotisation process, *Chem. Commun.*, 48(99), 12053–12055, 2012. Reproduced by permission of The Royal Society of Chemistry.)

2.2 PSMs OF MIL-100 FOR ADSORPTION AND CATALYSIS IN SOLUTION

The family of mesoporous MIL-100(M), M = Fe, Cr, Sc, Al, belongs to the isostructural MOFs (Wade and Dinca 2012). The MIL-100(Fe) MOF contains Fe_3O units with Fe(III) cations connected to the linkers of 1,3,5-benzenetricarboxylic (BTC), aka trimesic acid (Figure 2.10, Horcajada et al. 2007). Figure 2.10 in grayscale reproduces Figure 2 in color published in (Horcajada et al. 2007).

Table 2.3 shows MIL-100 MOFs modified by chemical adsorption and their respective applications in adsorption and heterogeneous catalysis in the liquid phase.

MIL-100(Cr) was modified with surface Brønsted acidic or basic groups and was then used for the adsorption of aromatic sulfur compounds and nitrogen heterocyclic compounds from model liquid fuels (Ahmed et al. 2013a). The PSM of MIL-100(Cr) was performed in solution at room temperature by grafting ethylenediamine (ED) as has been reported earlier (Hwang et al. 2008) for MIL-101(Cr). On the other hand, to synthesize its acidic counterpart AMSA–MIL-100(Cr), grafting of the acidic surface

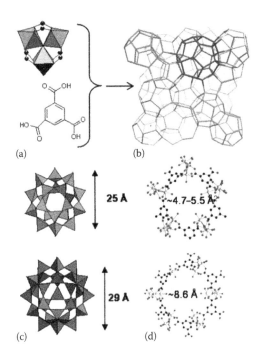

FIGURE 2.10 Structure of MIL-100(Fe). (a) A trimer of iron octahedra and trimesic acid. (b) Schematic view of one unit cell of MIL-100(Fe). (c) The two types of cages in the polyhedral mode. (d) Pentagonal and hexagonal windows in balls and sticks (Fe: grey; O: red; C: black). (From Horcajada, P., Surble, S., Serre, C., Hong, Do-Y., Seo, Y.-K., Chang, J.-S., Greneche, J.-M., Margiolaki, I., and Ferey, G., Synthesis and catalytic properties of MIL-100(Fe), an iron(III) carboxylate with large pores, *Chem. Commun.*, (27), 2820–2822, 2007. Reproduced by permission of The Royal Society of Chemistry.)

TABLE 2.3

Postsynthetic Modifications of MIL-100 via Adsorption

MOF	Modifier	Modified MOF	Application	Reference
MIL-100(Fe)(F)	$CuSO_4$ + Na citrate + glucose	Cu_2O/MIL-100(Fe)(F)	Desulfurization by adsorption	Khan and Jhung (2012)
MIL-100(Cr)(F)	Basic ED	ED–MIL-100(Cr)(F)	Desulfurization by adsorption	Ahmed et al. (2013a)
MIL-100(Cr)(F)	Acidic AMSA	AMSA–MIL-100(Cr)(F)	Desulfurization by adsorption	Ahmed et al. (2013a)
MIL-100(Cr)(F)	$CuCl_2$ + Na_2SO_3	CuCl/MIL-100(Cr)(F)	Denitrogenation by adsorption	Ahmed and Jhung (2014)
MIL-100(Fe)(F)	$AlCl_3$	$AlCl_3$/MIL-100(Fe)(F)	Denitrogenation by adsorption	Ahmed et al. (2014)
MIL-100(Fe)	H_2PdCl_4, PVP, ethanol, water	Pd@MIL-100(Fe)	Photocatalytic degradation	Liang et al. (2015)
MIL-100(Fe)	$HAuCl_4$, PVP, ethanol, water	Au@MIL-100(Fe)	Photocatalytic degradation	Liang et al. (2015)
MIL-100(Fe)	H_2PtCl_6, PVP, ethanol, water	Pt@MIL-100(Fe)	Photocatalytic degradation	Liang et al. (2015)
POM/MIL-100(Fe)	Ionic liquid, IL	IL/POM/MIL-100(Fe)	Esterification of oleic acid with ethanol to biodiesel	Wan et al. (2015)
MIL-100(Fe)	Heparin	Heparin/MIL-100(Fe)	Adsorption/release of caffeine in aqueous solutions	Bellido et al. (2015)

site was performed by the adsorption of aminomethanesulfonic acid (AMSA) onto the CUS of activated MIL-100(Cr) (Ahmed et al. 2013a).

The composite functional material MIL-100(Fe)/Cu(I) has been synthesized by chemical reduction of the complex of Cu(II) in water in the presence of MIL-100(Fe) (Khan and Jhung 2012). It is important to carefully control the oxidation state of copper, which could be either Cu(I) (a weak Lewis acid) or Cu(II) (a stronger Lewis acid). The presence of supported Cu in the Cu(I) oxidation state was demonstrated by x-ray photoelectron spectroscopy (XPS) analysis. In a similar way, MIL-100(Cr) was postmodified by CuCl to prepare CuCl/MIL-100(Cr) composite for the adsorptive removal of nitrogen aromatic compounds from model liquid fuels (Ahmed and Jhung 2014).

The strong Lewis acid $AlCl_3$ was adsorbed onto MIL-100(Fe)(F) in water solution, and the obtained composite $AlCl_3$/MIL-100(Fe)(F) was dried to remove water (Ahmed et al. 2014). The MOF-supported $AlCl_3$ has been shown to be effective in the adsorptive denitrogenation of liquid fuels in solution.

A nanocomposite material Pd@MIL-100(Fe) containing palladium metallic nanoparticles was fabricated via facile reduction of Pd^{2+} precursor by alcohol (Liang et al. 2015). The obtained Pd@MIL-100(Fe) exhibited strong photocatalytic activity toward degradation of the three representative pharmaceuticals and personal care products (PPCPs) in water, namely, theophylline, ibuprofen, and bisphenol A, under irradiation with visible light. The photocatalytic activity of Pd@MIL-100(Fe) was found to be much higher than that of MIL-100(Fe). Absorption of visible light is thought to occur in MIL-100(Fe), which has the suitable optical absorption spectrum and a distinct color, while supported metallic palladium nanoparticles constitute a common photocatalytic promoter among the metal-doped oxide photocatalysts. The nanoparticles of metallic palladium in Pd@MIL-100(Fe) are believed to assist in the separation of the photoexcited electron and hole formed in the MOF upon absorption of light.

The rather complex composite material has been prepared based on MIL-100(Fe) with both Lewis and Brønsted acidity (Wan et al. 2015). Here, the ionic liquid (IL) functionalized with the sulfonic acid group has been used in the PSM of polyoxometalate (POM)-containing MIL-100 through anion exchange (Figure 2.11).

The IL can diffuse into the interior of mesopores in MIL-100 and react with the heteropoly acid to form a heteropolyanion-based IL, which is believed to remain encapsulated within the mesocages. The sulfonic acid groups contributed to the available Brønsted acid sites, which work together with the Lewis acid sites supplied by the unsaturated coordinated metal centers Fe(III) and exhibit synergistic effects (Wan et al. 2015).

This is of interest to postsynthetically modify the mesoporous MOFs with biologically active molecules and use the obtained nanocomposites for emerging applications in biomedicine. Heparin is an anticoagulant (blood thinner) that prevents the formation of blood clots (Jones et al. 2011). The molecule of heparin contains several polar functional groups (sulfate, carboxylic, and hydroxyl) that are expected to strongly interact with the polar groups in the mesoporous MOFs such as the Fe(III) site in MIL-100(Fe).

Functionalization of the surface of MIL-100(Fe) with heparin has been reported (Bellido et al. 2015, Figure 2.12). In Figure 2.12, NP means "nanoparticle".

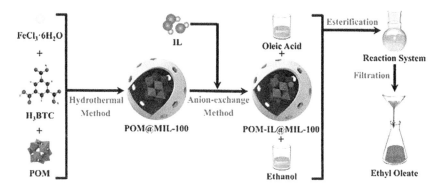

FIGURE 2.11 Synthesis of [SO$_3$H–(CH$_2$)$_3$–HIM]$_3$PW$_{12}$O$_{40}$@MIL-100. H$_3$BTC= benzene-1,3,5-tricarboxylic acid; IL= ionic liquid. (From Wan, H., Chen, C., Wu, Z.W., Que, Y.G., Feng, Y., Wang, W., Wang, L., Guan, G.F., and Liu, X.Q.: Encapsulation of heteropolyanion-based ionic liquid within the metal–organic framework MIL-100(Fe) for biodiesel production. *Chemcatchem.* 2015. 7(3). 441–449. Copyright Wiley-VCH Verlag GmbH & Co. KGaA. Reproduced with permission.)

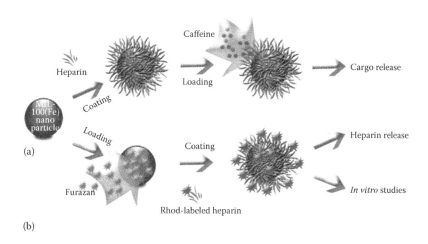

FIGURE 2.12 Schematic illustration of the formation of heparin-coated MIL-100(Fe) NPs loaded with specific cargoes via two differentiated pathways: (a) Heparin is selectively coated on the external surface of the NP followed by postloading with caffeine; and (b) preloading with furazan followed by external coating of the NP with Rhodamine-labeled heparin. The two systems are examined to evaluate the heparin content and robustness, the cargo loading–release profiles and to perform in vitro studies under simulated physiological conditions. (Adapted from Bellido, E., Hidalgo, T., Lozano, M.V., Guillevic, M., Simon-Vazquez, R., Santander-Ortega, M.J., Gonzalez-Fernandez, A., Serre, C., Alonso, M.J., and Horcajada, P.: Heparin-engineered mesoporous iron metal-organic framework nanoparticles: Toward stealth drug nanocarriers. *Adv. Healthcare Mater.* 2015. 4(8). 1246–1257. Copyright Wiley-VCH Verlag GmbH & Co. KGaA. Reproduced with permission.)

Furthermore, the adsorption of the additional biologically active compound on heparin-coated MIL-100(Fe) has been studied. The liporeductor caffeine has been selected as a model compound for small-molecule heterocyclic drugs, for proof-of-principle encapsulation on mesoporous MOFs that have been postsynthetically modified by adsorption of heparin. The encapsulation of caffeine by adsorption on both uncoated and heparin-coated nanoparticles of MIL-100(Fe) was conducted by simple impregnation in an aqueous solution. The encapsulated amounts of caffeine were comparable for the MIL-100(Fe) and the heparin/MIL-100(Fe) composite (Bellido et al. 2015).

2.3 PSMs OF MESOPOROUS MOFs OTHER THAN MIL-101 AND MIL-100 FOR APPLICATIONS BASED ON ADSORPTION IN SOLUTION

This is of interest to develop a suitable PSM for mesoporous MOFs other than MIL-100 and MIL-101 for applications involving adsorption/desorption in solution. Bio-MOF-100 is a mesoporous material that consists of metal–adeninate tetrahedral building blocks connected through biphenyldicarboxylate (BPDC) linkers (An et al. 2012). To prepare the azide-modified bio-MOF-100 (N_3-bio-MOF-100), biphenyldicarboxylic acid as the precursor of the linker was substituted with 2-azidobiphenyldicarboxylic acid. The azide-functionalized mesoporous bio-MOF-100, denoted N_3-bio-MOF-100, was postsynthetically modified (Liu et al. 2012), Figure 2.13. Figure 2.13 in grayscale reproduces Figure 1 in color published in (Liu et al. 2012).

The obtained postsynthetically modified bio-MOF-100 has been used as a platform for further PSMs with biomolecules, specifically di-L-phenylalanine peptides (Figure 2.14).

Thin films made with a mesoporous amino-functionalized MOF with the UiO-68 topology were grown on self-assembled monolayers (SAMs) deposited on gold (Hinterholzinger et al. 2012). These MOF films were postsynthetically covalently modified with the fluorescent dye Rhodamine B (Figure 2.15). The adsorbed Rhodamine B dye molecules are located inside the mesopores as was demonstrated through size-selective fluorescence quenching.

An interesting method for PSM termed "solvent-assisted ligand incorporation" (SALI) was developed to incorporate carboxylate-based functional groups into the Zr-based mesoporous MOF NU-1000 (Deria et al. 2013). The SALI method introduces the new functional groups to the MOF as the charge compensating strongly bound moieties attached to the Zr_6 node (Figure 2.16).

Perfluoroalkane carboxylates with chains of different lengths (C1–C9) were introduced at the Zr_6 nodes of NU-1000 MOF by SALI (Deria et al. 2013). The obtained fluoroalkane-functionalized mesoporous MOFs denoted SALI-n were studied as potential adsorbents for carbon dioxide. Such functionalized mesoporous MOFs could also be useful in the adsorption of organic compounds in solution.

The mesoporous bMOF-100 has a structure consisting of zinc–adeninate vertices interconnected through a bundle of three 4,4′-BPDC linkers (An et al. 2012).

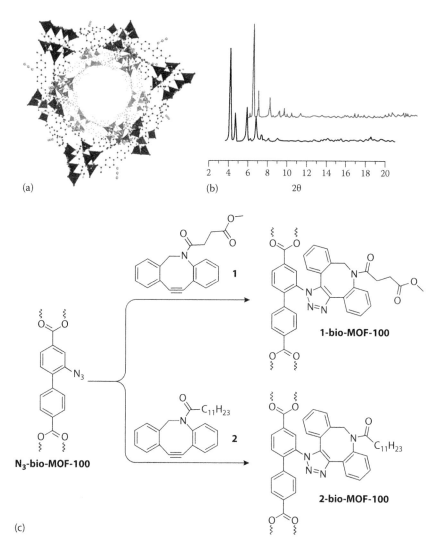

FIGURE 2.13 (a) Perspective view of an azide-decorated channel in N_3-bio-MOF-100. This image was generated from the single-crystal x-ray diffraction data for bio-MOF-100; Zn^{2+}, dark blue tetrahedra; O, dark red spheres; N, light blue spheres; C, dark gray spheres; H atoms omitted for clarity. (b) Powder x-ray diffraction (PXRD) pattern for bio-MOF-100 (black) and N_3-bio-MOF-100 (dark red). (c) Synthetic scheme for the strain promoted "click" modification of N_3-bio-MOF-100 with 1 and 2. (Adapted with permission from Liu, C., Li, T., and Rosi, N.L., Strain-promoted "click" modification of a mesoporous metal-organic framework, *J. Am. Chem. Soc.*, 134(46), 18886–18888. Copyright 2012 American Chemical Society.)

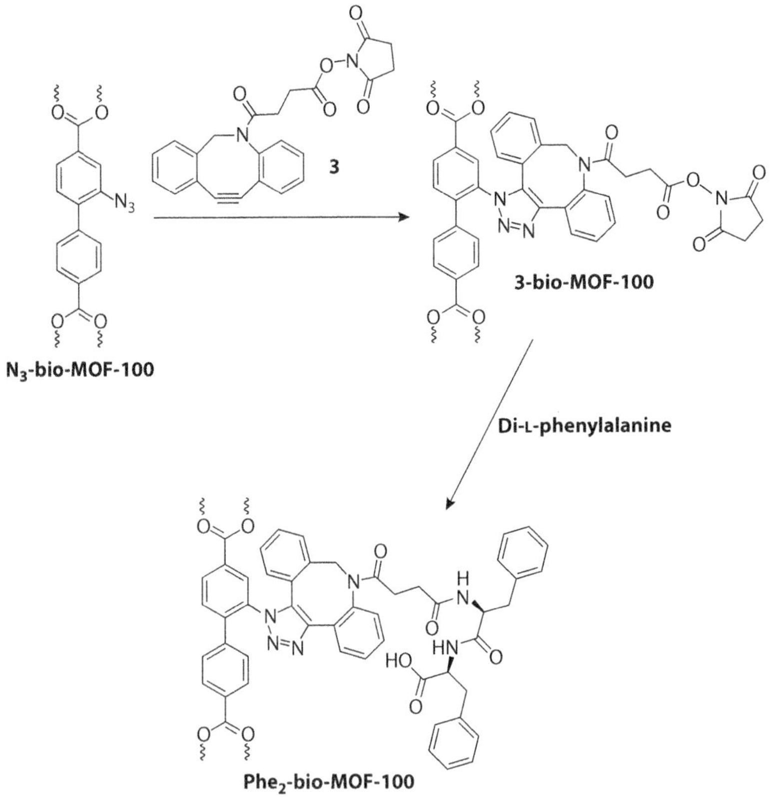

N₃-bio-MOF-100

3

3-bio-MOF-100

Di-L-phenylalanine

Phe₂-bio-MOF-100

FIGURE 2.14 Scheme for the strain-promoted "click" introduction of succinimidyl ester groups into bio-MOF-100 and the subsequent bioconjugation of the Phe2 peptide. (Adapted with permission from Liu, C., Li, T., and Rosi, N.L., Strain-promoted "click" modification of a mesoporous metal–organic framework, *J. Am. Chem. Soc.*, 134(46), 18886–18888. Copyright 2012 American Chemical Society.)

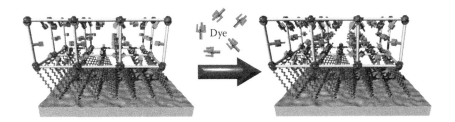

Dye

FIGURE 2.15 Schematic representation of oriented and functionalized MOF crystals grown on SAM-modified gold substrates (left) followed by the post-synthetic modification of the MOF thin film with dye molecules (right). (From Hinterholzinger, F.M., Wuttke, S., Roy, P., Preusse, T., Schaate, A., Behrens, P., Godt, A., and Bein, T., Highly oriented surface-growth and covalent dye labeling of mesoporous metal–organic frameworks, *Dalton Trans.*, 41(14), 3899–3901, 2012. Reproduced by permission of The Royal Society of Chemistry.)

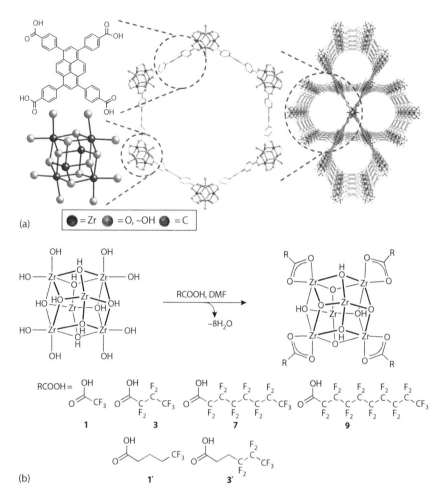

FIGURE 2.16 (a) Molecular representations of NU-1000 and (b) schematic representation of solvent-assisted ligand incorporation (SALI). (Reprinted with permission from Deria, P., Mondloch, J.E., Tylianakis, E., Ghosh, P., Bury, W., Snurr, R.Q., Hupp, J.T., and Farha, O.K., Perfluoroalkane functionalization of NU-1000 via solvent-assisted ligand incorporation: Synthesis and CO$_2$ adsorption studies, *J. Am. Chem. Soc.*, 135(45), 16801–16804. Copyright 2013 American Chemical Society.)

The postsynthetic exchange of ligand was developed to synthesize the series of functionalized mesoporous MOFs (Liu et al. 2015, Figure 2.17).

MOF DUT-32 has four different pores including mesopores (Grunker et al. 2014). An example of the PSMs of mesoporous MOFs based on deprotection of the linker is provided in a recent paper (Kutzscher et al. 2015). In the modified synthesis of DUT-32, 4,4′-biphenyldicarboxylic acid as a precursor of the linker was replaced with a protected (Boc)proline-functionalized unit, where Boc = N-(tert-butoxycarbonyl) protection group, resulting in a highly porous enantiomerically pure DUT-32-NHProBoc. However, the postsynthetic thermal deprotection of Boc–proline in this MOF resulted in racemization of the chiral center (Kutzscher et al. 2015).

The PSM of mesoporous MOFs for adsorption-related applications in solution is a promising approach to introduce certain adsorbate-specific functional groups that would cause an increase of the adsorption capacity and/or selectivity toward the target compounds. However, one of the limitations of the PSM of MIL-101 and MIL-100 is that the molecular size of introduced functional groups, beyond the simple functionalities such as sulfate or amino groups, could be quite large. Thus, PSM may cause a decrease in the mesopore size, leading to conversion of meso-porous MIL-101 and MIL-100 to microporous MOFs upon chemical modification. A radical approach to avoid this is to concentrate efforts on mesoporous MOFs from the other families, which are still rather scarce in the research on adsorption in solution.

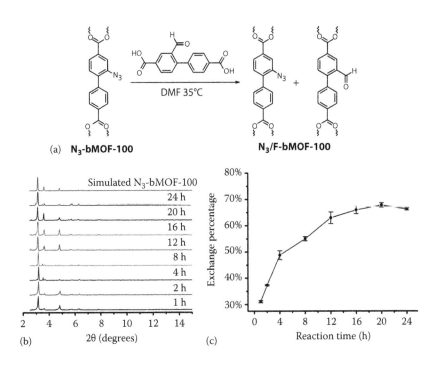

FIGURE 2.17 (a) Ligand-exchange reaction to produce N_3/F-bMOF-100. (b) PXRD comparing MOFs at different exchange time points with simulated pattern. (c) Percentage of F-BPDC in the product as determined by ^{1}H NMR.   (*Continued*)

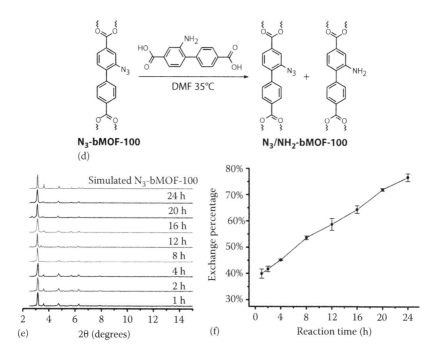

N₃-bMOF-100
(d)

N₃/NH₂-bMOF-100

(e)

(f)

FIGURE 2.17 (*Continued*) (d) Ligand-exchange reaction to produce N₃/NH₂-bMOF-100. (e) PXRD comparing MOFs at different exchange time points with simulated pattern. (f) Percentage of NH₂-BPDC in the product as determined by ¹H NMR. (Adapted with permission from Liu, C., Luo, T.Y., Feura, E.S., Zhang, C., and Rosi, N.L., Orthogonal ternary functionalization of a mesoporous metal–organic framework via sequential postsynthetic ligand exchange, *J. Am. Chem. Soc.*, 137(33), 10508–10511. Copyright 2015 American Chemical Society.)

REFERENCES

Ahmed, I., Z. Hasan, N. A. Khan, and S. H. Jhung. 2013a. Adsorptive denitrogenation of model fuels with porous metal-organic frameworks (MOFs): Effect of acidity and basicity of MOFs. *Applied Catalysis B: Environmental* 129:123–129.

Ahmed, I. and S. H. Jhung. 2014. Adsorptive denitrogenation of model fuel with CuCl-loaded metal-organic frameworks (MOFs). *Chemical Engineering Journal* 251:35–42.

Ahmed, I., J. W. Jun, B. K. Jung, and S. H. Jhung. 2014. Adsorptive denitrogenation of model fossil fuels with Lewis acid-loaded metal-organic frameworks (MOFs). *Chemical Engineering Journal* 255:623–629.

Ahmed, I., N. A. Khan, Z. Hasan, and S. H. Jhung. 2013b. Adsorptive denitrogenation of model fuels with porous metal-organic framework (MOF) MIL-101 impregnated with phosphotungstic acid: Effect of acid site inclusion. *Journal of Hazardous Materials* 250:37–44.

Ahmed, I., N. A. Khan, and S. H. Jhung. 2013c. Graphite oxide/metal-organic framework (MIL-101): Remarkable performance in the adsorptive denitrogenation of model fuels. *Inorganic Chemistry* 52(24):14155–14161.

Aijaz, A., A. Karkamkar, Y. J. Choi, N. Tsumori, E. Rönnebro, T. Autrey, H. Shioyama, and Q. Xu. 2012. Immobilizing highly catalytically active Pt nanoparticles inside the pores of metal–organic framework: A double solvents approach. *Journal of the American Chemical Society* 134(34):13926–13929.

An, J., O. K. Farha, J. T. Hupp, E. Pohl, J. I. Yeh, and N. L. Rosi. 2012. Metal-adeninate vertices for the construction of an exceptionally porous metal-organic framework. *Nature Communications* 3:Article number 604.

Bellido, E., T. Hidalgo, M. V. Lozano, M. Guillevic, R. Simon-Vazquez, M. J. Santander-Ortega, A. Gonzalez-Fernandez, C. Serre, M. J. Alonso, and P. Horcajada. 2015. Heparin-engineered mesoporous iron metal-organic framework nanoparticles: Toward stealth drug nanocarriers. *Advanced Healthcare Materials* 4(8):1246–1257.

Bernt, S., V. Guillerm, C. Serre, and N. Stock. 2011. Direct covalent post-synthetic chemical modification of Cr-MIL-101 using nitrating acid. *Chemical Communications* 47(10):2838–2840.

Bonnefoy, J., A. Legrand, E. A. Quadrelli, J. Canivet, and D. Farrusseng. 2015. Enantiopure peptide-functionalized metal–organic frameworks. *Journal of the American Chemical Society* 137(29):9409–9416.

Bromberg, L., Y. Klichko, E. P. Chang, S. Speakman, C. M. Straut, E. Wilusz, and T. A. Hatton. 2012. Alkylaminopyridine-modified aluminum aminoterephthalate metal-organic frameworks as components of reactive self-detoxifying materials. *ACS Applied Materials & Interfaces* 4(9):4595–4602.

Čendak, T., E. Žunkovič, T. U. Godec, M. Mazaj, N. Z. Logar, and G. Mali. 2014. Indomethacin embedded into MIL-101 frameworks: A solid-state NMR study. *Journal of Physical Chemistry C* 118(12):6140–6150.

Cohen, S. M. 2012. Postsynthetic methods for the functionalization of metal-organic frameworks. *Chemical Reviews* 112(2):970–1000.

Deria, P., J. E. Mondloch, O. Karagiaridi, W. Bury, J. T. Hupp, and O. K. Farha. 2014. Beyond post-synthesis modification: Evolution of metal-organic frameworks via building block replacement. *Chemical Society Reviews* 43(16):5896–5912.

Deria, P., J. E. Mondloch, E. Tylianakis, P. Ghosh, W. Bury, R. Q. Snurr, J. T. Hupp, and O. K. Farha. 2013. Perfluoroalkane functionalization of NU-1000 via solvent-assisted ligand incorporation: Synthesis and CO_2 adsorption studies. *Journal of the American Chemical Society* 135(45):16801–16804.

Férey, G., C. Mellot-Draznieks, C. Serre, and F. Millange. 2005a. Crystallized frameworks with giant pores: Are there limits to the possible? *Accounts of Chemical Research* 38(4):217–225.

Férey, G., C. Mellot-Draznieks, C. Serre, F. Millange, J. Dutour, S. Surblé, and I. Margiolaki. 2005b. A chromium terephthalate-based solid with unusually large pore volumes and surface area. *Science* 309(5743):2040–2042.

Goesten, M. G., J. Juan-Alcañiz, E. V. Ramos-Fernandez, K. B. Sai Sankar Gupta, E. Stavitski, H. van Bekkum, J. Gascon, and F. Kapteijn. 2011. Sulfation of metal–organic frameworks: Opportunities for acid catalysis and proton conductivity. *Journal of Catalysis* 281(1):177–187.

Grunker, R., V. Bon, P. Muller, U. Stoeck, S. Krause, U. Mueller, I. Senkovska, and S. Kaskel. 2014. A new metal-organic framework with ultra-high surface area. *Chemical Communications* 50(26):3450–3452.

Henschel, A., I. Senkovska, and S. Kaskel. 2011. Liquid-phase adsorption on metal-organic frameworks. *Adsorption* 17(1):219–226.

Hinterholzinger, F. M., S. Wuttke, P. Roy, T. Preusse, A. Schaate, P. Behrens, A. Godt, and T. Bein. 2012. Highly oriented surface-growth and covalent dye labeling of mesoporous metal-organic frameworks. *Dalton Transactions* 41(14):3899–3901.

Hong, D. Y., Y. K. Hwang, C. Serre, G. Ferey, and J. S. Chang. 2009. Porous chromium terephthalate MIL-101 with coordinatively unsaturated sites: Surface functionalization, encapsulation, sorption and catalysis. *Advanced Functional Materials* 19(10):1537–1552.

Horcajada, P., S. Surble, C. Serre, Do-Y. Hong, Y.-K. Seo, J.-S. Chang, J.-M. Greneche, I. Margiolaki, and G. Ferey. 2007. Synthesis and catalytic properties of MIL-100(Fe), an iron(III) carboxylate with large pores. *Chemical Communications* (27):2820–2822.

Huang, Y. B., Z. J. Lin, and R. Cao. 2011. Palladium nanoparticles encapsulated in a metal-organic framework as efficient heterogeneous catalysts for direct C_2 arylation of indoles. *Chemistry—A European Journal* 17(45):12706–12712.

Hwang, Y. K., D.-Y. Hong, J.-S. Chang, S. H. Jhung, Y.-K. Seo, J. Kim, A. Vimont, M. Daturi, C. Serre, and G. Férey. 2008. Amine grafting on coordinatively unsaturated metal centers of MOFs: Consequences for catalysis and metal encapsulation. *Angewandte Chemie International Edition* 47(22):4144–4148.

Jiang, D., L. L. Keenan, A. D. Burrows, and K. J. Edler. 2012. Synthesis and post-synthetic modification of MIL-101(Cr)-NH2 via a tandem diazotisation process. *Chemical Communications* 48(99):12053–12055.

Jones, C. J., S. Beni, J. F. K. Limtiaco, D. J. Langeslay, and C. K. Larive. 2011. Heparin characterization: Challenges and solutions. *Annual Review of Analytical Chemistry* 4(1):439–465.

Khan, N. A., Z. Hasan, and S. H. Jhung. 2014. Ionic liquids supported on metal-organic frameworks: Remarkable adsorbents for adsorptive desulfurization. *Chemistry—A European Journal* 20(2):376–380.

Khan, N. A. and S. H. Jhung. 2012. Low-temperature loading of Cu+ species over porous metal-organic frameworks (MOFs) and adsorptive desulfurization with Cu+-loaded MOFs. *Journal of Hazardous Materials* 237–238:180–185.

Kutzscher, C., H. C. Hoffmann, S. Krause, U. Stoeck, I. Senkovska, E. Brunner, and S. Kaskel. 2015. Proline functionalization of the mesoporous metal-organic framework DUT-32. *Inorganic Chemistry* 54(3):1003–1009.

Liang, R. W., S. G. Luo, F. F. Jing, L. J. Shen, N. Qin, and L. Wu. 2015. A simple strategy for fabrication of Pd@MIL-100(Fe) nanocomposite as a visible-light-driven photocatalyst for the treatment of pharmaceuticals and personal care products (PPCPs). *Applied Catalysis B—Environmental* 176:240–248.

Liu, C., T. Li, and N. L. Rosi. 2012. Strain-promoted "click" modification of a mesoporous metal-organic framework. *Journal of the American Chemical Society* 134 (46):18886–18888.

Liu, C., T. Y. Luo, E. S. Feura, C. Zhang, and N. L. Rosi. 2015. Orthogonal ternary functionalization of a mesoporous metal-organic framework via sequential postsynthetic ligand exchange. *Journal of the American Chemical Society* 137(33):10508–10511.

Mitchell, L., B. Gonzalez-Santiago, J. P. S. Mowat, M. E. Gunn, P. Williamson, N. Acerbi, M. L. Clarke, and P. A. Wright. 2013. Remarkable Lewis acid catalytic performance of the scandium trimesate metal organic framework MIL-100(Sc) for C-C and C=N bond-forming reactions. *Catalysis Science & Technology* 3(3):606–617.

Modrow, A., D. Zargarani, R. Herges, and N. Stock. 2012. Introducing a photo-switchable azo-functionality inside Cr-MIL-101-NH$_2$ by covalent post-synthetic modification. *Dalton Transactions* 41(28):8690–8696.

Serra-Crespo, P., E. V. Ramos-Fernandez, J. Gascon, and F. Kapteijn. 2011. Synthesis and characterization of an amino functionalized MIL-101(Al): Separation and catalytic properties. *Chemistry of Materials* 23(10):2565–2572.

Tanabe, K. K., and S. M. Cohen. 2011. Postsynthetic modification of metal-organic frameworks—A progress report. *Chemical Society Reviews* 40(2):498–519.

Wade, C. R. and M. Dinca. 2012. Investigation of the synthesis, activation, and isosteric heats of CO_2 adsorption of the isostructural series of metal-organic frameworks M3(BTC)2 (M = Cr, Fe, Ni, Cu, Mo, Ru). *Dalton Transactions* 41(26):7931–7938.

Wan, H., C. Chen, Z. W. Wu, Y. G. Que, Y. Feng, W. Wang, L. Wang, G. F. Guan, and X. Q. Liu. 2015. Encapsulation of heteropolyanion-based ionic liquid within the metal-organic framework MIL-100(Fe) for biodiesel production. *Chemcatchem* 7(3):441–449.

Wang, Z. Y., G. Li, and Z. G. Sun. 2013. Denitrogenation through adsorption to sulfonated metal-organic frameworks. *Acta Physico-Chimica Sinica* 29(11):2422–2428.

Wittmann, T., R. Siegel, N. Reimer, W. Milius, N. Stock, and J. Senker. 2015. Enhancing the water stability of Al-MIL-101-NH_2 via postsynthetic modification. *Chemistry—A European Journal* 21(1):314–323.

Yadav, M., A. Aijaz, and Q. Xu. 2012. Highly catalytically active palladium nanoparticles incorporated inside metal-organic framework pores by double solvents method. *Functional Materials Letters* 5(4):1250039 (4pp.).

3 Mechanistic Studies of Activation of Mesoporous MOFs

After the synthesis and purification of metal–organic frameworks (MOFs) including mesoporous MOFs, it is usually necessary to perform the so-called "sorbent activation" procedure, before conducting the adsorption experiment in the solution or in the gas phase. The term "sorbent activation" commonly describes two kinds of chemical reactions to be conducted with the as-prepared MOF.

The first kind of activation involves removing an excess of precursors of the linker or the metal that remain in the obtained powder of the MOF after its chemical synthesis. The standard purification procedure of the "as-synthesized" mesoporous MOF normally includes washing the raw product with a suitable solvent to remove the excess reactants. Highly polar solvents such as dimethylformamide (DMF), ethanol, and methanol are normally used. However, the washing procedure is often not sufficient to fully remove organic contaminants; therefore, subsequent thermal treatment of the MOF is required. Depending upon the stability of the particular MOF to air, thermal activation is conducted in ambient air or, more often, in vacuum. In case of any doubt about the propensity of the MOF toward oxidation, activation in vacuum under moderate heating is normally conducted.

The second kind of activation involves changes at the metal node site in the MOF. In many cases, chemical synthesis of the MOF includes water or another highly polar liquid compound, for example, DMF, to be used as the solvent. As a result, the "as-synthesized" MOF routinely contains water or DMF molecules adsorbed via the coordination bonds to the metal cation site or to the linker. The metal site in the MOF with the coordinated solvent molecule is usually not active in adsorption; thus, the so-called "coordinatively unsaturated site" (CUS) of the metal is needed. Unless the metal site is activated and CUS is present, the adsorption capacity and/or the selectivity of the MOF are often very low. The need for generating metal CUS in mesoporous MOFs was realized early on. Virtually every paper on mesoporous MOFs (and on other kinds of MOFs) includes the "sorbent activation" procedure usually described in the Experimental part. In quite many cases, the activation of the MOF for adsorption-based applications is considered to be a routine procedure not worth dedicated research. In those cases, the previously published activation procedure is simply reproduced under further research work.

In a limited number of academic papers, the importance of learning about the molecular-level details of chemical and structural changes at the CUS and/or linker site upon or during the activation procedure has been acknowledged. Then, research has been devoted specifically to learning about the transformations of the MOF during and upon the activation procedure itself. In such papers, the MOF sorbent is characterized by two sets of methods of instrumental chemical and structural analysis: (1) "routine" MOF characterization to check its chemical composition and structure, and (2) characterization methods specifically chosen to be applied directly to the MOF sample under the activation process, that is, in situ. The characterization methodologies applied in the studies of the MOF activation procedure can be classified as the ex situ and in situ methods. In this chapter, representative publications on the studies of the activation procedure of mesoporous MOFs are described.

The ex situ characterization of the MOF in the activation procedure is conducted by applying temperature, vacuum, and/or the other methods and then by removing the activated MOF from the activation apparatus to be analyzed. This is usually undertaken to check the integrity of the framework after the thermal treatment, and powder x-ray diffraction (XRD) is the standard method. The loss of mass in the MOF during the activation procedure is also conveniently studied using the thermogravimetric analysis (TGA).

On the other hand, in the in situ characterization of the MOF sorbent as conducted directly during the activation procedure, the sorbent activation chamber needs to be equipped with suitable scientific instrument. In situ characterization of the MOF under the activation procedure has been conducted by thermodiffractometric analysis, aka thermo-XRD, using the XRD instrument equipped with thermal attachment and optional vacuum specimen chamber. Given that mesoporous MOFs have their most characteristic XRD patterns at low diffraction angles, in situ small-angle x-ray scattering (SAXS) and in situ small-angle XRD instruments would be needed.

MIL-100(Fe) is one of the well-investigated mesoporous MOFs. MIL-100(Fe) is particularly interesting for mechanistic studies of adsorption and catalysis, since it combines the following properties. First, the high surface area and pore volume make it suitable for applications in adsorption in solution, catalysis, encapsulation, and the release of biomedically active compounds.

Second, Fe(III) is relatively nontoxic as compared to Cr(III), which is present in the other common MOF, MIL-101(Cr). Third, the originally present Fe(III) CUS may change its valence state to Fe(II) during the activation procedure, and this new valence state of the CUS can be useful in catalytic oxidation reactions (Table 3.1). Fourth, commercially available mesoporous MOF F300 Basolite, which is similar to MIL-100(Fe) (Dhakshinamoorthy et al. 2012, Sanchez-Sanchez et al. 2015), has been introduced to the market, and it has been studied for adsorption in solution (Centrone et al. 2011, Dai et al. 2014).

In the early study of thermal activation of MIL-100(Fe)(F), TGA indicated the three weight losses in the temperature range of 298–873 K (Horcajada et al. 2007). Prior to this TGA analysis, MIL-100(Fe)(F) was washed with water to remove the excess of the 1,3,5-benzenetricarboxylic (BTC) linker (free trimesic acid) and dried at room temperature. The first loss of mass at ca. 40.1% at 373 K was attributed to desorption of the "free" water molecules located inside the pores. The second loss at 473 K (~4.5%) was due to water molecules interacting with Fe_3O trimers; this loss can be assigned to "chemisorbed water." Finally, significant weight loss of ca. 35.3% at 573 K was found to be due to the thermal decomposition of trimesic acid (Horcajada et al. 2007) as the framework-building material which has ultimately resulted in destruction of the MOF.

TABLE 3.1
Studies of Activation of MIL-100 and Similar MOFs for Adsorption and Catalysis Applications

MOF	MOF Activation	Method(s) of MOF Characterization and Learning about the Activation	Changes in the MOF during/upon Activation	Reference
MIL-100(Fe)(F)	Dried at RT	TGA at 298–873 K	Loss of physisorbed and chemisorbed water; destruction of BTC linker	Horcajada et al. (2007)
MIL-100(Fe)(F)	100°C in nitrogen	In situ FTIR with CO adsorption; in situ small-angle XRD; Mössbauer spectra	Partial reduction of Fe(III) to Fe(II) at > 150°C; framework is stable at < 250°C	Yoon et al. (2010)
MIL-100(Fe)(F)	—	Thermo-XRD; adsorption of CO versus CO_2, propane versus propene and propyne, pyridine; in situ FTIR; microcalorimetry	Framework is stable at < 280°C; π-backdonation from Fe(II) CUS to adsorbates with C=C and C=O bonds	Leclerc et al. (2011)
Basolite F300	293 K in vacuum	XANES, EXAFS, XRD	No reduction of Fe(III) to Fe(II)	Sciortino et al. (2015)
MIL-100(Al)		Elemental analysis, thermo-XRD, TGA, FTIR; solid-state NMR at 130°C–350°C	Removal of adsorbed water; strongly trapped H_3BTC precursor	Haouas et al. (2011)
MIL-100(Al)	573 K for 5 h	298–623 K; FTIR, adsorption of CO, pyridine, acetonitrile	Al(III) CUS detected by CO adsorption	Volkringer et al. (2012)

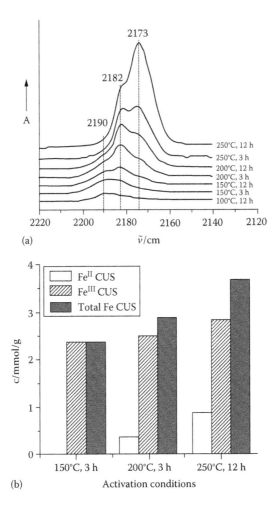

FIGURE 3.1 (a) Infrared (IR) spectra of MIL-100 under a stream of 10% CO at 25°C after activation under a helium flux at various temperatures over 3 or 12 h. (b) Amount of Fe^{III} CUS and Fe^{II} CUS detected by IR analysis upon CO adsorption at −173°C on MIL-100(Fe) activated under high vacuum at different temperatures. (From Yoon, J.W., Seo, Y.K., Hwang, Y.K., Chang, J.S., Leclerc, H., Wuttke, S., Bazin, P. et al.: Controlled reducibility of a metal–organic framework with coordinatively unsaturated sites for preferential gas sorption. *Angew. Chem. Int. Ed.* 2010. 49(34). 5949–5952. Copyright Wiley-VCH Verlag GmbH & Co. KGaA. Adapted with permission.)

The particular advantage of mesoporous MOF with iron CUS is that the interchange between the two major stable oxidation states of iron, Fe(III) and Fe(II), becomes possible. Controlled reduction of Fe(III) CUS in MIL-100(Fe)(F) has been studied by in situ Fourier transform infrared (FTIR) spectroscopy during thermal activation of this MOF (Yoon et al. 2010). In situ FTIR analysis utilized carbon monoxide as a probe molecule to learn about the oxidation states of the metal site in MIL-100(Fe)(F)

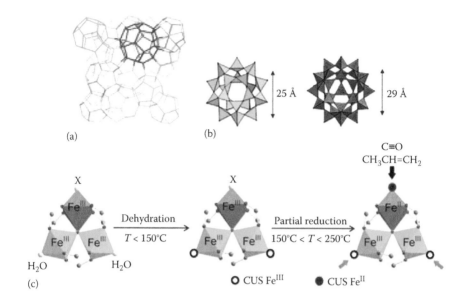

FIGURE 3.2 Representations of MIL-100(Fe): (a) one unit cell, (b) two types of mesoporous cages shown as polyhedra, and (c) formation of Fe^{III} CUS and Fe^{II} CUS in an octahedral iron trimer of MIL-100(Fe) by dehydration and partial reduction from the departure of anionic ligands (X^- = F^- or OH^-). (From Yoon, J.W., Seo, Y.K., Hwang, Y.K., Chang, J.S., Leclerc, H., Wuttke, S., Bazin, P. et al.: Controlled reducibility of a metal–organic framework with coordinatively unsaturated sites for preferential gas sorption. *Angew. Chem. Int. Ed.* 2010. 49(34). 5949–5952. Copyright Wiley-VCH Verlag GmbH & Co. KGaA. Reproduced with permission.)

(Figure 3.1). At room temperature, the CO molecule does not strongly interact with Fe(III) sites, as judged by the weak absorbance band at $2190 \ cm^{-1}$.

Furthermore, gradual increase in the activation temperature in the presence of carbon monoxide in the helium carrier gas resulted in the appearance of two new absorption bands of MIL-100(Fe) at 2182 and $2173 \ cm^{-1}$ (Yoon et al. 2010). These spectral bands remained after treatment of the sample in vacuum, and they were assigned to the CO molecules adsorbed on Fe(II) CUS, which are in situ generated by the thermally driven reduction of Fe(III) sites. The resultant change in the oxidation state of Fe(III) CUS is shown in Figure 3.2.

The intensity of these IR bands increased with an increase in the activation temperature and with the duration of thermal treatment. To determine the number of thermally generated Fe(II) CUS, the sample was cooled to −173°C and the value of the molar absorption coefficient for the vibration ν(CO) at 2.7 cm/μmol was utilized. The integrity of the mesoporous framework in MIL-100(Fe)(F) during heating to temperatures up to 250°C was confirmed by in situ small-angle XRD in vacuum (Yoon et al. 2010).

The Fe(II) CUS is a softer Lewis acid than the Fe(III) site, so thermally reduced MIL-100(Fe) containing Fe(II) has been prepared and studied for the selective adsorption of heteroaromatic compounds versus aromatic hydrocarbons in solution

(Maes et al. 2011). While "regular" MIL-100(Fe)(F) demonstrated zero adsorption capacity for the aromatic sulfur compounds benzothiophene (BT) and dibenzothiophene (DBT) in the hydrocarbon solvent, partial reduction with formation of Fe(II) CUS resulted in the uptake of 9 wt.% BT and 11 wt.% DBT. Fe(II) as the soft Lewis acid is believed to selectively interact with the heteroaromatic π-electron system of aromatic sulfur compounds, which are soft Lewis bases.

In the following study (Leclerc et al. 2011), the adsorption of small molecules in the gas phase (CO, CO_2, propane, propene, and propyne) was investigated on partially reduced MIL-100(Fe)(F), with the goal of assessing the mechanism of adsorption. The proposed adsorption via the π-backdonation mechanism would be useful in the selective adsorption of olefins and aromatic and heteroaromatic compounds. Formation of CUS with the Lewis acid upon removal of the adsorbed water at 150°C in vacuum was demonstrated by the adsorption of probe molecules CO and pyridine. Compared to previous work (Yoon et al. 2010), higher activation temperatures were studied in detail (Leclerc et al. 2011). Thermal activation at 250°C results in the formation of ca. 850 μmol/g Fe(II) CUS or 17% of the total number of Fe sites. Activation at the higher temperature of 280°C leads to collapse of the MIL-100 structure. The presence of Fe(II) CUS results in the interaction with molecules, which are sensitive to the π-backdonation. The d-electron of the reduced Fe(II) site is transferred to an antibonding π^* orbital of molecules with one double and triple bonds such as CO, propene, and propyne. If the adsorbate has no isolated double or triple bond (propane, pyridine, and CO_2), partial reduction to Fe(II) does not result in the π-backdonation effect, as judged by the adsorption enthalpy determined by microcalorimetric data.

Commercially available mesoporous MOF F300 Basolite is manufactured by Badische Anilin und Soda Fabrik (BASF) and exclusively sold by Sigma-Aldrich. F300 Basolite is commonly termed FeBTC, and it is structurally similar to MIL-100(Fe) (Dhakshinamoorthy et al. 2012), but F300 Basolite contains only one kind of the mesopore as determined by nitrogen adsorption (Sanchez-Sanchez et al. 2015). Small-angle XRD measurements of Basolite F300 show its simpler structure compared to MIL-100(Fe), as well as the lower crystallinity of the former (Sanchez-Sanchez et al. 2015). The oxidation state of Fe in Basolite F300 after several treatments has been studied by x-ray absorption near-edge structure (XANES) and extended x-ray absorption fine structure (EXAFS) spectroscopy (Sciortino et al. 2015). Specifically, the samples have been (1) the "as-received" material from Sigma-Aldrich, (2) the FeBTC after 24 h of exposure to ambient air at room temperature termed "24 h air," and (3) activated FeBTC after being kept at 393 K in vacuum for 3 h to remove adsorbed water. The XANES data of the Fe site in all three forms of Basolite F300 indicate the Fe^{3+} form, based on measurements of the pairs of reference compounds FeO versus Fe_2O_3 and $FeSO_4$ versus $Fe_2(SO_4)_3$. Based on the spectroscopic data, it has been proposed that the building block of F300 MOF closely resembles iron(III) acetate. The building block of F300 has been assigned the formula $Fe_3(\mu_3\text{-}O)(\mu\text{-}OAc)_6(H_2O)_3$, where Ac is the CH_3CO group. This structural "acetate model" contains three equivalent iron atoms, with each of them bound to six oxygen ligands in the octahedral geometry. After thermal activation, the structure was found to only slightly relax keeping the building block of the "acetate model" intact (Figure 3.3).

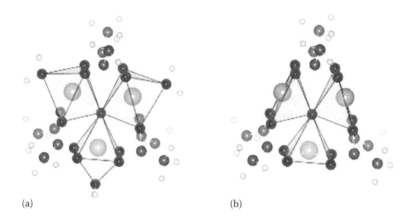

(a)   (b)

FIGURE 3.3 Acetate model picture where iron atoms are inside the polyhedra, oxygen atoms are red, carbon atoms are brown, and hydrogen atoms are white. (a) The acetate model built to fit the EXAFS and (b) the dehydrated acetate model. (Adapted with permission from Sciortino, L., Alessi, A., Messina, F., Buscarino, G., and Gelardi, F.M., Structure of the FeBTC metal-organic framework: A model based on the local environment study, *J. Phys. Chem. C*, 119(14), 7826–7830, 2015. Copyright 2015 American Chemical Society.)

Besides the Lewis acid properties of its Fe(III) CUS, Basolite F300 MOF is known to possess higher activity in the acid-catalyzed reactions than MIL-100(Fe) (Dhakshinamoorthy et al. 2012), apparently due to the Brønsted acid sites in the former. It has been proposed (Sciortino et al. 2015) that the Brønsted acid sites in F300 are constituted of terminal carboxyl groups, which interrupt the structural order of the network. This would result in Basolite F300 having lower crystallinity compared to MIL-100; indeed, this hypothesis has been confirmed by XRD measurements (Sciortino et al. 2015).

Aluminum-containing mesoporous MOFs are attractive for health care and environmental applications, due to very low toxicity of aluminum. In MIL-100(Al), each of the Al(III) ions of the Al_3O cluster is connected to the carboxylate groups of the BTC linker, thus forming the octahedra, as in the case of the well-investigated MIL-100(Cr), and the terminal position is occupied either by a water molecule or by an OH^- group. The structural changes upon thermally induced removal and the subsequent adsorption of water molecule on MIL-100(Al) were investigated by solid-state nuclear magnetic resonance (NMR) spectroscopy (Haouas et al. 2011). The "as-prepared" MIL-100(Al) was activated by a two-step process. First, the solid was treated in anhydrous DMF at 150°C, in order to dissolve the extra-framework organic species entrapped within the pores. Then, the obtained powder was treated in water under reflux overnight. The "as-prepared" MIL-100(Al) contains a significant amount of extra-framework trimesic acid H_3BTC in the pores; most of this H_3BTC precursor can be removed upon activation with DMF/water. However, about 0.3–0.5 molecules of the H_3BTC precursor, on average, strongly interact with each Al_3 trimer in the framework, so their complete removal is difficult (Haouas et al. 2011, Figure 3.4).

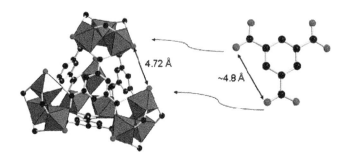

FIGURE 3.4 Possible interaction mode of extra-framework H_3BTC with the trimers Al_3 in MIL-100(Al). (Reprinted with permission from Haouas, M., Volkringer, C., Loiseau, T., Férey, G., and Taulelle, F., Monitoring the activation process of the giant pore MIL-100(Al) by solid-state NMR, *J. Phys. Chem. C*, 115(36), 17934–17944, 2011. Copyright 2011 American Chemical Society.)

Overall, MIL-100(Al) has shown thermal stability up to 370°C, with rather small structural changes leading to the lower symmetry of the framework. At 350°C, one water molecule per Al_3 trimer was found to desorb from this trimer (Haouas et al. 2011).

After hydrothermal synthesis of MIL-100(Al), the obtained yellowish powder was washed with water and dried at room temperature in air (Volkringer et al. 2012). Then, this powder was dehydrated in DMF in a Teflon-lined autoclave at 423 K for 5 h. Finally, the obtained white powder was calcined at 573 K for 5 h to remove encapsulated DMF, resulting in activated MIL-100(Al). Infrared spectra of MIL-100(Al) have been recorded after evacuation at the variable temperature between room temperature and 623 K. The adsorbed water was observed via the combination bands at ca. 5300 cm⁻¹. In addition, several impurities such as carboxylic acid and nitrates that remain from the synthesis were observed. The nitrate species apparently originated from the synthesis of MIL-100(Al), which involves the addition of nitric acid. In the experiments with CO adsorption, the major band $\nu(CO)$ found at 2195–2184 cm⁻¹ originates in the interaction of CO molecule with large amount of Al(III) CUS formed upon desorption of water. On the other hand, weak $\nu(CO)$ band at 2210 cm⁻¹ is attributed to Al(III) defects with high Lewis acidity. The Brønsted acidity of MIL-100(Al) was found to mainly originate from coordinated water in the noncompletely dehydrated sample, rather than from the Al–OH groups in the framework (Volkringer et al. 2012).

The MOF activation procedure is important in achieving the desired adsorption characteristics, especially for adsorption in the nonaqueous solvents, where the CUS is expected to remain intact and to interact directly with the adsorbate molecules. Even for nominally the same MOF to be used as the sorbent, the differences in the activation procedure as conducted by different research labs could lead to substantially different adsorption performance. This makes it hard to draw valuable correlations between the identity of the adsorbate and the adsorption behavior of the given MOF. It would be highly desirable to establish "standard" protocols for the activation of each mesoporous MOF, in order to be able to compare the performance of the sorbent as reported by different research labs.

REFERENCES

Centrone, A., E. E. Santiso, and T. A. Hatton. 2011. Separation of chemical reaction intermediates by metal–organic frameworks. *Small* 7(16):2356–2364.

Dai, J., M. L. McKee, and A. Samokhvalov. 2014. Adsorption of naphthalene and indole on F300 MOF in liquid phase by the complementary spectroscopic, kinetic and DFT studies. *Journal of Porous Materials* 21(5):709–727.

Dhakshinamoorthy, A., M. Alvaro, P. Horcajada, E. Gibson, M. Vishnuvarthan, A. Vimont, J. M. Greneche, C. Serre, M. Daturi, and H. Garcia. 2012. Comparison of porous iron trimesates Basolite F300 and MIL-100(Fe) as heterogeneous catalysts for Lewis acid and oxidation reactions: Roles of structural defects and stability. *ACS Catalysis* 2(10):2060–2065.

Haouas, M., C. Volkringer, T. Loiseau, G. Férey, and F. Taulelle. 2011. Monitoring the activation process of the giant pore MIL-100(Al) by solid state NMR. *Journal of Physical Chemistry C* 115(36):17934–17944.

Horcajada, P., S. Surble, C. Serre, Do-Y. Hong, Y.-K. Seo, J.-S. Chang, J.-M. Greneche, I. Margiolaki, and G. Ferey. 2007. Synthesis and catalytic properties of MIL-100(Fe), an iron(iii) carboxylate with large pores. *Chemical Communications* (27):2820–2822.

Leclerc, H., A. Vimont, J. C. Lavalley, M. Daturi, A. D. Wiersum, P. L. Llwellyn, P. Horcajada, G. Ferey, and C. Serre. 2011. Infrared study of the influence of reducible iron(III) metal sites on the adsorption of CO, CO_2, propane, propene and propyne in the mesoporous metal-organic framework MIL-100. *Physical Chemistry Chemical Physics* 13(24):11748–11756.

Maes, M., M. Trekels, M. Boulhout, S. Schouteden, F. Vermoortele, L. Alaerts, D. Heurtaux et al. 2011. Selective removal of N-heterocyclic aromatic contaminants from fuels by Lewis acidic metal-organic frameworks. *Angewandte Chemie—International Edition* 50(18):4210–4214.

Sanchez-Sanchez, M., I. de Asua, D. Ruano, and K. Diaz. 2015. Direct synthesis, structural features, and enhanced catalytic activity of the Basolite F300-like semiamorphous Fe-BTC framework. *Crystal Growth & Design* 15(9):4498–4506.

Sciortino, L., A. Alessi, F. Messina, G. Buscarino, and F. M. Gelardi. 2015. Structure of the FeBTC metal-organic framework: A model based on the local environment study. *Journal of Physical Chemistry C* 119(14):7826–7830.

Volkringer, C., H. Leclerc, J. C. Lavalley, T. Loiseau, G. Ferey, M. Daturi, and A. Vimont. 2012. Infrared spectroscopy investigation of the acid sites in the metal-organic framework aluminum trimesate MIL-100(Al). *Journal of Physical Chemistry C* 116(9):5710–5719.

Yoon, J. W., Y. K. Seo, Y. K. Hwang, J. S. Chang, H. Leclerc, S. Wuttke, P. Bazin et al. 2010. Controlled reducibility of a metal-organic framework with coordinatively unsaturated sites for preferential gas sorption. *Angewandte Chemie—International Edition* 49(34):5949–5952.

4 Stability of Mesoporous MOFs in Water

A majority of both microporous and mesoporous metal–organic frameworks (MOFs) are not stable in water due to the hydrolysis of their coordination "node," which contains the complex of the metal ion and organic linker. However, several well-studied mesoporous MOFs, particularly those from the MIL-100 and MIL-101 families, show rather high stability to water, in either the liquid or the vapor phase. This stability has offered opportunities for quite many applications of water-stable mesoporous MOFs, in particular in environmental and biomedical fields.

4.1 WATER STABILITY OF MIL-101(Cr)

Several papers have been published in recent years to test the stability of the as-synthesized mesoporous MOFs from the MIL-101 family (Table 4.1).

The stability of MIL-101(Cr) to steam-containing nitrogen and water vapors has been tested using the high throughput method with variable temperature and humidity (Low et al. 2009). The most severe conditions under which structural loss of MIL-101(Cr) was not observed were in steam with water content at 50 mol.% at 325°C. The steady loss of low angle reflections in x-ray diffraction (XRD) indicated the loss of mesoporosity. In addition, MIL-101(Cr) changed its characteristic green color to brown due to the presence of Cr(III) sites, apparently due to chemical decomposition (Low et al. 2009). The water physisorption properties and the stability of MIL-101(Cr) were studied in deionized water at 323 K for 24 h, and MIL-101(Cr) was found to be stable (Kusgens et al. 2009). The hydrothermal stability of MIL-101(Cr) was studied by mixing it with deionized water and heating in a closed vessel under boiling at 100°C, without stirring, for up to 7 days (Hong et al. 2009). MIL-101(Cr) was found to be stable, as was demonstrated by the XRD and N_2 adsorption isotherms before and after treatment with water.

The adsorption of water vapor on MIL-101(Cr) was found to be up to 1 g water/g sorbent at 40°C–140°C under a water vapor pressure of 5.6 kPa (Ehrenmann et al. 2011). MIL-101(Cr) was found to be stable over several cycles of water adsorption/desorption, as was determined by the water and nitrogen adsorption/desorption isotherms. Based on the literature available by 2011, the authors (Ehrenmann et al. 2011) concluded that MIL-101(Cr) was the most promising MOF for heat transformation applications that use water adsorption/desorption, such as thermally driven heat pumps and adsorption chillers.

TABLE 4.1

Stability of Pure MIL-101(Cr) in Water

MOF	Liquid/Vapor?	Testing Conditions	Characterization	Result	Reference
MIL-101(Cr)(F)	Vapor	Variable temperature, RH	XRD	Stable < 325°C	Low et al. (2009)
MIL-101(Cr)(F)	Liquid	323 K	XRD, N_2 sorption, H_2O sorption	Stable for 24 h	Kusgens et al. (2009)
MIL-101(Cr)(F)	Liquid	100°C	XRD and N_2 sorption	Stable for 7 days	Hong et al. (2009)
MIL-101(Cr)(F)	Vapor	40°C–140°C	H_2O and N_2 sorption	Stable	Ehrenmann et al. (2011)
MIL-101(Cr)(F)	Liquid	RT	XRD, ζ-potential	Stable at pH = 2–10; unstable at pH ≥ 12	Chen et al. (2012)
MIL-101(Cr)(F); MIL-101(Cr)	Vapor	100°C 50 mbar water	XRD, TGA, gas sorption	Stable	Liang et al. (2013)
MIL-101(Cr)(F)	Vapor	Adsorption isotherms, simulations	N_2 and water sorption, XRD	Stable	De Lange et al. (2013)

The stability of MIL-101(Cr) in water at variable pH has been studied by measuring the ζ-potential (Chen et al. 2012) as needed for understanding the adsorption of organic dyes in water. The positive ζ-potential of MIL-101(Cr) increased moderately with increase in pH, and it subsequently decreased slightly at pH > 7. This indicates the stability of MIL-101(Cr) toward hydrolysis in the approximate range pH = 2–12. However, at pH > 12, a sharp decrease of the ζ-potential and its negative numeric values indicates the decomposition of MIL-101(Cr), as was confirmed by complementary XRD data (Chen et al. 2012).

MIL-101(Cr) was prepared by the new "HF-free" method, characterized by the XRD, thermogravimetric analysis (TGA), and adsorption isotherms of water, CO_2, CH_4, and N_2 in comparison with the "conventional" MIL-101(Cr) prepared using hydrofluoric acid (Liang et al. 2013). The structure of both "HF-free" MIL-101(Cr) and "conventional" MIL-101(Cr)(F) remained intact after exposure to water vapor at 100°C and at partial pressure 50 mbar water vapor (Liang et al. 2013).

The adsorption of water and methanol vapors on MIL-101(Cr)(F) and MIL-101(Cr) was studied by a combination of experiments and computer simulations (De Lange et al. 2013). The difference in the adsorption isotherms at small water loadings was found to be due to the different partial charge on Cr(III). The O atom of the $Cr_3O–(OH)$ cluster of MIL-101(Cr) forms hydrogen bonds with water. At very small water loadings, the heat of adsorption drops quickly from ~80 kJ/mol to the value just above the heat of evaporation of water, when the loading is slightly increased. This indicates that water–water interactions control the adsorption, after the Cr(III) sites have been occupied with adsorbate molecules (De Lange et al. 2013).

4.2 WATER STABILITY OF CHEMICALLY MODIFIED MIL-101(Cr)

Significant attention has been paid to improving water stability and the water adsorption capacity of MIL-101 via postsynthetic modifications (PSMs), as well as via the introduction of suitable functional groups by chemical synthesis from modified precursors (Table 4.2).

New mesoporous MIL-101(Cr)–SO_3H with strong Brønsted acid sites on the pore surfaces was synthesized from chromium(VI) oxide, monosodium 2-sulfoterephthalic acid, and hydrochloric acid in water (Akiyama et al. 2011). Elemental analysis has shown that MIL-101(Cr)–SO_3H does not contain chloride ions. This finding is in contrast to the case of MIL-101(Cr), which contains F$^-$ ions after synthesis using hydrofluoric acid. When immersed in water, MIL-101(Cr)–SO_3H produces an acidic solution that can be titrated with 0.1 M NaOH to determine the concentration of surface sulfonic acid groups. MIL-101(Cr)–SO_3H has been found to be stable in boiling water for >24 h, and no change in the powder XRD or nitrogen adsorption isotherm was found.

Following this approach, the same group reported the synthesis of four MIL-101(Cr) with different polar substituents in the BDC linker (–H, –NO_2, –NH_2, –SO_3H) from the respective substituted BDC acids (Akiyama et al. 2012). The MOFs were

TABLE 4.2
Stability of Chemically Modified MIL-101(Cr) in Water

MOF	PSM?	Water		Testing	Stable?	Reference
MIL-101(Cr)–SO₃H	No	Liquid		100°C, >24 h	Yes	Akiyama et al. (2011)
MIL-101(Cr)–X		Vapor		25°C	Yes	Akiyama et al. (2012)
X = –H	No					
–NO₂	No					
–NH₂	Reduction of –NO₂					
–SO₃H	No					
MIL-101(Cr)–DAAP	PSM with DAAP in toluene	Liquid		RT, water/acetonitrile, 24 h at pH 10	Yes	Wang et al. (2013)
MIL-101(Cr)	No	Vapor		40°C–140°C	40 cycles	Khutia et al. (2013)
MIL-101(Cr)–NO₂	HNO₃/H₂SO₄					
MIL-101(Cr)–NH₂	Reduction with SnCl₂					
MIL-101(Cr)(F)	Resorcinol–formaldehyde resin	Vapor		Adsorbs 0.8 g/g water	Yes	Wickenheisser et al. (2015)

Abbreviation: DAAP, dialkylaminopyridines.

characterized by synchrotron powder x-ray diffraction (PXRD), nitrogen adsorption isotherms, and the Brunauer–Emmett–Teller (BET) total surface area measurements. Water vapor sorption properties of these MOFs were investigated using the water adsorption and desorption isotherms and TGA. All tested MOFs adsorbed large amounts of water in the range 0.8–1.2 g/g. The adsorbed water can be desorbed at relatively mild conditions, for example, at 5–100 kPa and <373 K. The hydrophobic parameters of the $-NH_2$, $-SO_3H$, and $-NO_2$ groups (Hansch et al. 1973) were found to correlate in the water adsorption in MIL-101(Cr) with the respective substituents in the linker (Akiyama et al. 2012). Thus, the water adsorption profile can be tuned by the systematic change of the substituents in the linker of the MOF.

"Fluorine-free" MIL-101(Cr) synthesized as reported in (Bromberg et al. 2012) was modified by complexation with dialkylaminopyridines (DAAPs, Wang et al. 2013; Figure 4.1). In Figure 4.1, reference [32] is (Férey et al. 2005) which reported synthesis and structure of "conventional" fluoride-containing MIL-101(Cr) often denoted MIL-101(Cr)(F).

Specifically, the DAAP units were incorporated into the mesopores of MIL-101(Cr) by complexation with a Cr(III) coordinatively unsaturated site (CUS) in thermally activated MOF by the reaction of the respective DAAPs in anhydrous toluene. Upon this PSM, no destruction of the MIL-101 framework was observed. The DAAP–MOF composite was found to be catalytically active in the hydrolysis of organophosphorus ester paraoxon in water/acetonitrile mixtures under ambient conditions (Figure 4.2).

The conversion approached 100% after 24 h at pH = 10. This implies the stability of the DAAP–MOF catalyst in water under the described conditions (Wang et al. 2013).

Amino and nitro groups–functionalized MIL-101(Cr) have been synthesized by the PSM of HF-free MIL-101(Cr) (Khutia et al. 2013). Specifically, hydrophilic $-NO_2$ and $-NH_2$ groups were introduced into the BDC linker of MIL-101(Cr) in order to achieve water vapor adsorption at low partial pressures. Aminated MIL-101(Cr)–NH_2 showed the highest water loadings at ca. 1.0 g H_2O/g sorbent. After 40 cycles of water adsorption/desorption, powder XRD and the BET total surface area were used to check the structural integrity of the MOFs. For MIL-101(Cr) with the $-NH_2$ groups, the BET surface area decreased by 6.3%, while for MIL-101(Cr) with $-NO_2$ the surface area decreased by 25% or 20%, depending on the number of $-NO_2$ groups in the MOF.

For emerging industrial applications, it is important to convert "as-prepared" mesoporous MOFs in the form of powders into the desired shapes. Recently, the synthesis of monolithic MOF composites was reported using MIL-101(Cr) with resorcinol–formaldehyde–based xerogel as the binding agent (Wickenheisser et al. 2015). The polymer monoliths could be loaded with up to 77 wt.% MIL-101(Cr)(F), the total surface area and porosities of MIL-101(Cr) were retained as found by the nitrogen adsorption tests, and the reported MOF-containing monoliths were found to be mechanically stable. These monoliths adsorbed amounts of water vapor close to the calculated values based on the weight content of MIL-101. These hydrophilic, MOF-containing composite materials can be used for heat transformation applications, for example, thermally driven adsorption chillers or adsorption heat pumps.

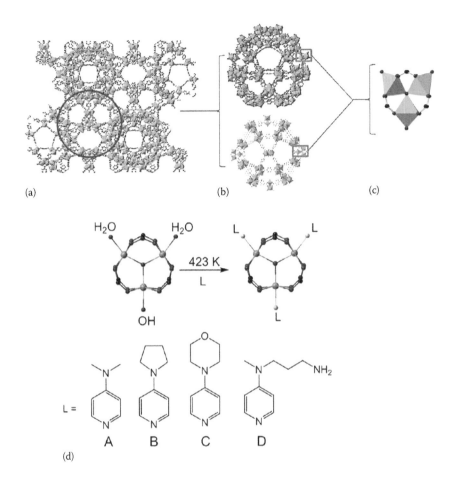

FIGURE 4.1 Schematic representation of (a) the unit cell of MIL-101 crystal structure with the boundaries of two types of cages (large cage shown in pink, and small cage shown as a blue circle), (b) polyhedral model view of the two cages; (c) the trimeric building block sharing an oxygen center and chelated by six carboxylates. Chromium octahedra, oxygen, and carbon atoms are in green, red, and blue, respectively. (d) Evolution of unsaturated metal sites of MIL-101 after vacuum heating, and postmodification of dehydrated MIL-101 through coordination of DAAP-type molecules onto its chromium unsaturated sites. The unit structure depicted corresponds to the published structure of chromium terephthalate MIL-101 [32] with formula $Cr_3OH(H_2O)_2O[(O_2C-C_6H_4-(CO_2)]_3$, wherein fluoride is substituted with –OH, as we utilized HF-free synthesis of the MOF [29]. The presence of three ligands (L) per unit is shown, which corresponds to the structures of MIL-101-A, MIL-101-B, and MIL-101-C (Table 4.1), based on elemental analyses. Crystal structures depicted were built utilizing published lattice parameters [29,32] and applying Crystal Maker software. (Reprinted with permission from Wang, S., Bromberg, L., Schreuder-Gibson, H., and Hatton, T.A., Organophophorous ester degradation by chromium(III) terephthalate metal–organic framework (MIL-101) chelated to N,N-dimethylaminopyridine and related aminopyridines, *ACS Appl. Mater. Interfaces*, 5(4), 1269–1278. Copyright 2013 American Chemical Society.)

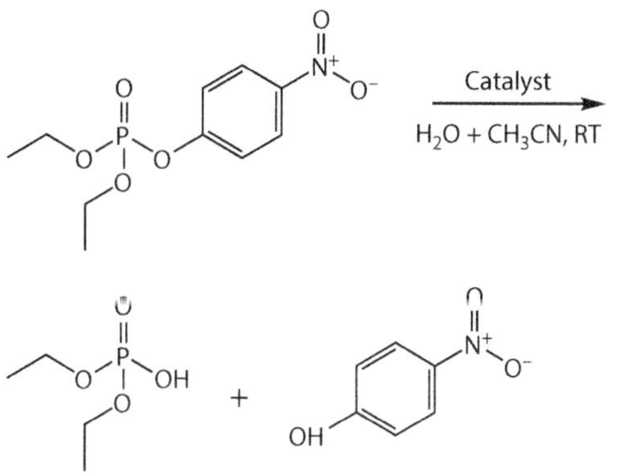

FIGURE 4.2 Catalytic hydrolysis of paraoxon with new MOF composites. (Adapted with permission from Wang, S., Bromberg, L., Schreuder-Gibson, H., and Hatton, T.A., Organophophorous ester degradation by chromium(III) terephthalate metal-organic framework (MIL-101) chelated to *N,N*-dimethylaminopyridine and related aminopyridines, *ACS Appl. Mater. Interfaces*, 5(4), 1269–1278. Copyright 2013 American Chemical Society.)

4.3 WATER STABILITY OF MIL-101(Fe), MIL-101(Al), AND MIL-101(V)

Table 4.3 shows published data on the water stability of MIL-101 other than MIL-101(Cr).

Iron is less toxic than chromium, yet there are only few studies of the water stability of MIL-101(Fe). The nanocrystalline MOF NMOF MIL-101(Fe)–NH_2 quickly dissolved in the water solution of Na_4EDTA at 37°C, and it was also found unstable in phosphate-buffered saline (PBS) solution (Taylor-Pashow et al. 2009). MIL-101(Fe)–NH_2 was prepared from the dimethyl aminoterephthalate precursor (Figure 4.3) as the active component of composite proton-conductive membranes (Wu et al. 2013).

The obtained MIL-101(Fe)–NH_2 was found to be stable in liquid water at room temperature (RT) for 24 h. The mechanism of proton transfer in the obtained mixed-matrix membranes (MMMs) containing MIL-101(Fe)–NH_2 is proposed (Wu et al. 2013) as shown in Figure 4.4. The proton H^+ is believed to be delivered to the surface of the MOF and stabilized via hydrogen bonding and is subsequently transferred to the next MOF crystal by water as the carrier.

Aluminum is a nontoxic metal, and MIL-101(Al) would be suitable for environmental (Khan et al. 2013) and health care (Horcajada et al. 2012) applications. The water stability of MIL-101(Al) has not been reported, while it has been studied for the adsorption of organic dye from water (Haque et al. 2014). MIL-101(Al)–NH_2 is usually synthesized in dimethylformamide (DMF); in water, it quickly transforms (Stavitski et al. 2011) to microporous MIL-53(Al)–NH_2 (Figure 4.5).

The reported stability of MIL-101(Al)–NH_2 in liquid water at <50°C for 12 h (Haque et al. 2014) has been challenged by the more recent finding (Wittmann et al. 2015)

TABLE 4.3
Stability of MIL-101 Other Than MIL-101(Cr) in Water

MOF	Postmodification	Liquid/Vapor?	Liquid Phase	Testing	Sorbent Characterization	Stability	Reference
MIL-101(Fe)–NH$_2$	No	Liquid	PBS or EDTA in water	37°C	XRD	Unstable	Taylor-Pashow et al. (2009)
MIL-101(Fe)–NH$_2$	No	Liquid	Water	RT, 24 h.	XRD	Stable	Wu et al. (2013)
MIL-101(Al)–NH$_2$	No	Liquid	Water	<50°C	XRD	Stable for 12 h	Haque et al. (2014)
MIL-101(Al)–NH$_2$	No	Liquid	Water	RT	XRD, ^{1}H, ^{27}Al and ^{13}C NMR, ATR–FTIR, AAS	<5 min	Wittmann et al. (2015)
MIL-101(Al)–URPh	Phenylisocyanate	Liquid	Water	RT	Above	Stable 7 days	Wittmann et al. (2015)
MIL-101(V)	No	Vapor	—	—	Adsorption CO$_2$, CH$_4$, N$_2$, O$_2$, water	Water protects V(III)	Yang et al. (2015)

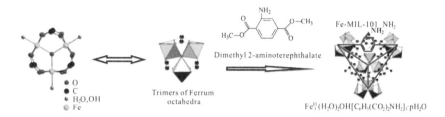

FIGURE 4.3 Schematic 3D representation of the tetrahedra built up from trimers of ferrum octahedra and 2-amino-1,4-benzenedicarboxylate moieties in Fe-MIL-101–NH₂. (From Wu et al. (2013). Reproduced by permission of The Royal Society of Chemistry.)

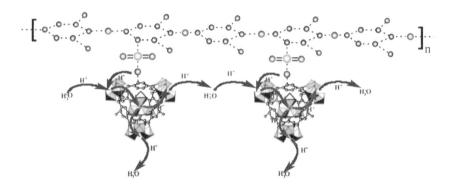

FIGURE 4.4 Schematic illustration of the possible pathway of proton transfer in the membranes. S, O, C, and N are represented as yellow, red, gray, and blue, respectively; H atoms are omitted. (From Wu et al. (2013). Reproduced by permission of The Royal Society of Chemistry.)

that MIL-101(Al)–NH₂ undergoes transformation to MIL-53(Al)–NH₂ in water at RT in <5 min. The PSM of microporous MIL-53(Al)–NH₂ results in moisture-resistant MOFs (Nguyen and Cohen 2010). The same approach has been applied to mesoporous MOFs; the –NH₂ group of the mesoporous MIL-101(Al)–NH₂ was modified with phenylisocyanate C₆H₅–NCO, and the obtained hydrophobic MIL-101(Al)–URPh was found to be stable in liquid water for 7 days (Wittmann et al. 2015).

Mesoporous MIL-101(V) has been synthesized from vanadium trichloride (VCl₃) and terephthalic acid in DMF at 423 K for 24 h, washed with DMF, and dried at RT under argon (Yang et al. 2015). Protection from ambient air was needed in order to avoid oxidation of V(III) CUS to V(IV). The adsorption of water vapor on MIL-101(V) was assessed to increase the stability of this MOF to oxidation with air. Oxygen adsorption tests were conducted for activated MIL-101(V) and MIL-101(V)(H₂O) as obtained after pre-adsorption of water vapors. O₂ adsorption has been found to be fully reversible, and it has been concluded that O₂ cannot strongly bind on the V(III) site upon pre-adsorption of water vapor. Experimental

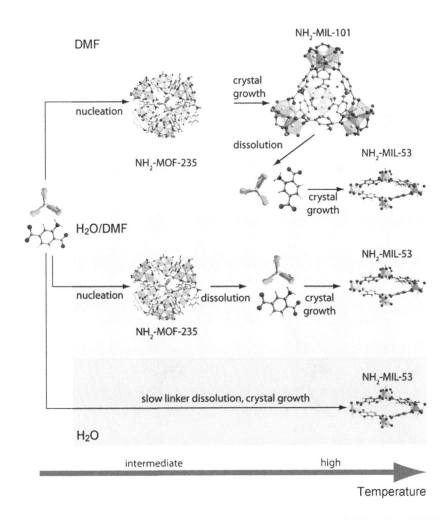

FIGURE 4.5 The sequence of events during the crystallization of terephthalate-based MOFs in different media: low precursor concentrations (DMF) and high precursor concentrations (H₂O/DMF or H₂O). C, gray; H, white; N, blue; O, red; Al, yellow; Cl, green. (From Stavitski, E., Goesten, M., Juan-Alcañiz, J., Martinez-Joaristi, A., Serra-Crespo, P., Petukhov, A.V., Gascon, J., and Kapteijn, F.: Kinetic control of metal–organic framework crystallization investigated by time-resolved in situ x-ray scattering. *Angew. Chem. Int. Ed.* 2011. 50(41). 9624–9628. Copyright Wiley-VCH Verlag GmbH & Co. KGaA. Reproduced with permission.)

data were complemented with DFT calculations of the binding energy of oxygen and water. The binding energy of one molecule of O_2 to the open metal V(III) site was 93.278 kJ/mol, which decreased to 26.5 kJ/mol after one H_2O molecule was pre-adsorbed on the site. When the "protective" water layer was removed, the x-ray photoelectron spectroscopy (XPS) showed oxidation of V(III) towards V(IV).

4.4   WATER STABILITY OF MIL-100(Fe)

Table 4.4 shows reported data on the water stability of nonmodified MIL-100(Fe).

Water physisorption properties and the stability of MIL-100(Fe)(F) were studied in deionized water at 323 K for 24 h (Kusgens et al. 2009); MIL-100(Fe)(F) was found to be stable.

The fluorine-free colloids of nanocrystalline MIL-100(Fe), termed nanocrystalline MOFs or NMOFs, namely, NMOF MIL-100(Fe), have been found to be stable in liquid water at 37°C for 3 days (Agostoni et al. 2013). However, NMOF MIL-100(Fe) lost 23%–27% of its linker in the water solutions of tris and PBS, due to the interactions of Fe(III) with phosphate ions.

Fluorine-free NMOF MIL-100(Fe) was prepared by microwave-assisted hydrothermal synthesis from Fe(III) chloride and trimesic acid (Bellido et al. 2014). With MIL-100(Fe) NMOF in water, the pH decreased to 3.1 due to deprotonation of the 1,3,5-benzenetricarboxylic (BTC) linker, and agglomeration occurred. The subsequent treatment of MIL-100(Fe) NMOF in a 0.1 M KF solution in water resulted in a 50% increase of the surface area; the negative ζ-potential indicated that the F$^-$ anion coordinates with Fe(III). The KF-treated MIL-100(Fe) NMOF formed stable and monodispersed colloids in simulated gastric fluid (SGF; 0.137 M HCl, pH =1.2) and in simulated intestinal fluid (SIF; KH_2PO_4 and NaOH at pH 6.8). In acidic medium at pH ~1.2, leaching of Fe(III) and BTC acid from MIL-100(Fe) to solution occurred.

The fluoride-free MIL-100(Fe) suspended in deionized water at 22°C caused a decrease in the pH to 2.9 due to hydrolysis of the MOF (Bezverkhyy et al. 2016). This finding stresses the importance of careful control of the pH in studies of water stability of the MOFs. Then, neutral conditions were maintained in the suspension at pH = 7 by adding sodium hydroxide, and the framework of MIL-100(Fe) collapsed, forming poorly crystallized ferrihydrite. Furthermore, the stability of MIL-100(Fe) was also checked by reflux in water at 100°C. After reflux, the XRD pattern did not change; however, a strong decrease of the BET total surface area was observed. The XRD demonstrated the presence of hydrolysis products, namely, nanoparticles of α-Fe_2O_3 (Bezverkhyy et al. 2016).

4.5   WATER STABILITY OF MIL-100(Cr)

Table 4.5 shows published data on the water stability of MIL-100(Cr).

It is well-known that the "standard" method for the synthesis of MIL-100(Cr) is in the presence of hydrofluoric acid (HF) in water solution (Férey et al. 2004). In a study in 2010, activated MIL-100(Cr)(F) was found to be stable in pure water at RT after as long as 12 months, based on powder XRD (Cychosz and Matzger 2010). To check the effects of strong inorganic acid used in the synthesis, MIL-100(Cr)(F), MIL-100(Cr)(Cl), and MIL-100(Cr)(SO_4) were prepared in the presence of hydrofluoric acid, hydrochloric acid, and sulfuric acid, respectively, and their ability to adsorb water vapor was investigated (Akiyama et al. 2010). All three MOFs adsorbed large amounts of water from vapor, up to 0.6 g water/g at the rather moderate humidity $p/p_0 < 0.6$. It has been concluded that the partial pressure of water at which adsorption of water occurs is controlled by the counter-anions F, Cl, or SO_4, probably due to their

TABLE 4.4

Stability of Pure MIL-100(Fe) in Water

Liquid/Vapor?	Liquid Phase	Conditions	Method	Result	Reference
Liquid	Water	323 K, 24 h	XRD, N_2 sorption, H_2O sorption	Stable	Kusgens et al. (2009)
Vapor	—	40°C–140°C, $p/p_0 < 0.4$	XRD, H_2O sorption	Stable 40 heating/cooling cycles	Jeremias et al. (2012)
Liquid	Water	RT	BET, N_2 sorption, TEM, FTIR, TGA, XRD, GI-WAXS, FESEM, ζ-potential	Stable	Marquez et al. (2012)
Liquid	Water; tris; PBS	RT	XRD, UV-Vis, Monte Carlo simulations	Stable water 3 days; unstable tris or PBS	Agostoni et al. (2013)
Liquid	PBS, SGF, SIF	37°C	XRD, FTIR, N_2 adsorption, TGA, DLS, ζ-potential	Unstable 24 h	Bellido et al. (2014)
Liquid	Water	100°C; 22°C at pH = 7	XRD, N_2 adsorption, thermal analysis, TEM, SEM, FTIR	Spontaneous acidification; unstable	Bezverkhyy et al. (2016)

TABLE 4.5

Stability of Pure MIL-100(Cr) in Water

MOF	Liquid/Vapor?	Liquid Phase	Conditions	Method	Result	Reference
MIL-100(Cr)(F)	Liquid	Water	25°C, 1 year	XRD	Stable	Cychosz and Matzger (2010)
MIL-100(Cr)(F); MIL-100(Cr)(Cl); MIL-100(Cr)(SO$_4$)	Vapor	—	298 K	XRD, H$_2$O sorption, TGA	Stable	Akiyama et al. (2010)
MIL-100(Cr), fluoride-free	Liquid	Water	RT	BET, N$_2$ sorption, TEM, FTIR, TGA, XRD, GI-WAXS, FESEM, ζ-potential	Unstable, leading to pH = 2.5	Marquez et al. (2012)
MIL-100(Cr)(F)	Vapor	—	Adsorption isotherms, simulations	N$_2$ and water sorption, XRD	Stable	De Lange et al. (2013)

different hydration energies. MIL-100(Cr)(F), MIL-100(Cr)(Cl), and MIL-100(Cr) (SO$_4$) released the adsorbed water at 353 K, as was found by TGA. After as many as 2000 successive water adsorption/desorption tests, the "used" MIL-100(Cr)(F) showed the same adsorption capacity as the freshly prepared material (Akiyama et al. 2010).

In new fluorine-free methods of synthesis, MIL-100(Cr) was prepared by the microwave-assisted hydrothermal method, with the following activation in ethanol (Marquez et al. 2012). MIL-100(Cr) was found to form a stable colloidal solution in ethanol and methanol; however, in pure water, large particles were formed leading to a nonstable colloid, with phase separation occurring in a few minutes. In contrast to ethanol where the positive values of the ζ-potential were found, the suspension of MIL-100(Cr) in water has shown negative ζ-potential. In addition, an acidic solution was formed with pH = 2.5 after suspending MIL-100(Cr) in water (Marquez et al. 2012).

Adsorption of water vapor has been compared on MIL-100(Cr)(F) and MIL-101(Cr)(F) (De Lange et al. 2013) by experiment and computer simulations. For MIL-100(Cr), the heat of adsorption decreases more gradually than for MIL-101(Cr) as the water loading increases. The conclusion was that the mesocavities in MIL-100(Cr) are smaller and that fewer water molecules are adsorbed at saturation; thus, adsorbate/sorbent interactions are more significant than those in MIL-101(Cr).

4.6 WATER STABILITY OF MIL-100(Al)

Table 4.6 shows the data on the water stability of nonmodified MIL-100(Al).

MIL-100(Al) prepared in water has been found to be stable only within the narrow range 0.5 < pH < 0.7 (Volkringer et al. 2009). MIL-100(Al) demonstrated the same shape of the water adsorption isotherm as mesoporous MIL-100(Fe) (Jeremias et al. 2012). The adsorption/desorption of water vapor on MIL-100(Fe) and MIL-100(Al) was studied at temperature cycling from 40°C to 140°C, at the small relative pressures p/p(0) < 0.4. Adsorption/desorption proceeds with a rather small hysteresis, and up to 0.75 g of water per gram of sorbent can be reversibly adsorbed. MIL-100(Al) has shown the lower adsorbed amount of water than MIL-100(Fe), consistent with the smaller pore volume. The stability of MOFs was retained after as many as 40 temperature cycles with water adsorption and desorption (Jeremias et al. 2012).

MIL-100(Al) was reported to be isostructural to MIL-100(Fe) and to have mesopores (Marquez et al. 2012). However, during synthesis of the former, microporous aluminum trimesate MIL-96(Al) was also produced as a major by-product. To bypass the water instability of MIL-100(Al), methyl ester of BDC was used as a precursor, and the obtained MIL-100(Al) formed a stable colloid in water (Marquez et al. 2012).

The hydrothermal stability of MIL-100(Al) was tested by water adsorption in comparison with that of microporous MIL-96(Al) and MIL-110(Al) (Čelič et al. 2013). MIL-100(Al) has shown higher stability than MIL-110(Al) in water. The water adsorption capacity of MIL-100(Al) achieved 0.7 g water per gram of sorbent.

Textural properties of aluminum trimesate MOFs synthesized in the presence of cationic surfactant cetyltrimethylammonium bromide (CTAB) at 120°C were

TABLE 4.6
Stability of Pure MIL-100(Al) in Water

Liquid/Vapor?	Conditions	Method	Result	Reference
Liquid	210°C, 72 h	XRD, SEM, ^{27}Al MAS NMR, TGA	Stable only at pH = 0.5–0.7	Volkringer et al. (2009)
Vapor	40°C–140°C, $p/p_0 < 0.4$	XRD, H_2O sorption	Stable 40 heating/cooling cycles	Jeremias et al. (2012)
Liquid	RT	BET, N_2 sorption, TEM, FTIR, TGA, XRD, GI-WAXS, FESEM, ζ-potential	Stable when synthesized from methyl ester of BDC	Marquez et al. (2012)
Vapor	40°C–140°C at 75% RH	Water sorption isotherms, XRD	Stable 40 cycles; water sorption capacity 0.7 g/g	Čelič et al. (2013)
Liquid	RT	XRD, BET, FTIR	Stable at pH = 2.3–2.5	Seoane et al. (2015)

systematically studied (Seoane et al. 2015). Depending on the pH and composition of the mixture of solvents (water + ethanol EtOH), MOFs of the three topologies were obtained: MIL-96, mesoporous MIL-100, and MIL-110. Mesoporous MIL-100(Al) with crystallite sizes 30 ± 10 nm were obtained at a molar ratio $H_2O/EtOH = 3.4$ at pH 2.6. As in the case of previously reported synthesis of MIL-100(Al) (Volkringer et al. 2009), the synthesis of MIL-100(Al) in the presence of a structural template is still limited, with a very narrow pH range (Figure 4.6).

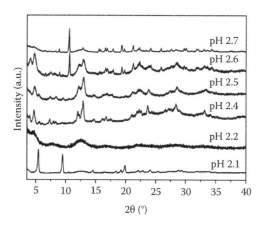

FIGURE 4.6   XRD patterns of the samples obtained at different pH with CTAB/Al and H_2O/EtOH molar ratios of 0.6 and 3.4, respectively. (From Seoane, B., Dikhtiarenko, A., Mayoral, A., Tellez, C., Coronas, J., Kapteijn, F., and Gascon, J., Metal organic framework synthesis in the presence of surfactants: Towards hierarchical MOFs?, *CrystEngComm*, 17(7), 1693–1700, 2015. Reproduced by permission of The Royal Society of Chemistry. Creative Commons Attribution 3.0 Unreported License.)

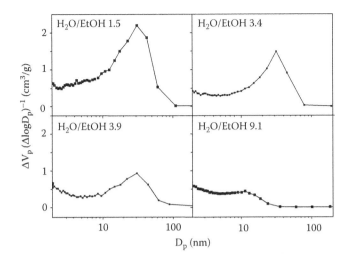

FIGURE 4.7 BJH pore size distribution curves of the samples synthesized at pH 2.5 in mixtures of H_2O and EtOH with different molar ratios and a CTAB/Al ratio of 0.6. (From Seoane, B., Dikhtiarenko, A., Mayoral, A., Tellez, C., Coronas, J., Kapteijn, F., and Gascon, J., Metal organic framework synthesis in the presence of surfactants: Towards hierarchical MOFs?, *CrystEngComm*, 17(7), 1693–1700, 2015. Reproduced by permission of The Royal Society of Chemistry. Creative Commons Attribution 3.0 Unreported License.)

At both the higher and the lower pH, microporous reaction products MIL-96 and MIL-110 were formed (Seoane et al. 2015). The obtained MIL-100(Al) shows a hierarchical porosity that includes microporosity and nonordered mesopores between the MOF nanoparticles; the maximum of the pore size distribution of this MIL-100 could be varied between 3 and 33 nm (Figure 4.7).

4.7 WATER STABILITY OF CHEMICALLY MODIFIED MIL-100

Chemically modified MIL-100 sorbents have been studied much less compared to modified MIL-101. Activated MIL-100(Cr) has been postsynthetically modified by "grafting" with ethylene glycol (EG), diethylene glycol (DEG), triethylene glycol (TEG), or ethylenediamine (EN) (Wickenheisser et al. 2013), resulting in MIL-100(Cr)-EG, MIL-100(Cr)-DEG, and MIL-100(Cr)-EN, respectively. This surface grafting resulted in a decrease of the BET total surface areas and pore volume. The water adsorption isotherms of grafted MIL-100(Cr)–EG, MIL-100(Cr)–DEG, and MIL-100(Cr)–EN show higher adsorbed amount of water, compared to nonmodified MIL-100(Cr). MIL-100(Cr)–EN showed the highest stability for water vapors. The samples of MIL-100(Cr)–EN retained water adsorption capacity after exposure to a humidified gas flow and cycling the temperature 20 times between 140°C and 40°C under a constant partial vapor pressure of water (Wickenheisser et al. 2013).

Isostructural MOFs with metal cations other than Fe(III), Cr(III), and Al(III) were investigated much less. Mesoporous MIL-100(V) was synthesized from

VCl$_3$ and H$_3$BTC in water in a Teflon autoclave at 473 K for 72 h, washed with water and dried at RT under argon, to yield the hydrated form, MIL-100(V) (H$_2$O) (Yang et al. 2015). Therefore, MIL-100(V) is apparently unstable in air due to oxidation of the V(III) site. As in the case of MIL-101(V) (Yang et al. 2015), adsorption of molecular oxygen on MIL-100(V) was found to be reversible when the CUS was protected by a pre-adsorbed water molecule. Mesoporous MIL-100(Sc) has been synthesized and utilized as a Lewis catalyst in organic reactions (Mitchell et al. 2013); however, its water stability has not been studied.

4.8 WATER STABILITY OF MESOPOROUS MOFs OTHER THAN MIL-101 AND MIL-100

Table 4.7 shows published data on water adsorption studies on mesoporous MOFs other than MIL-101 and MIL-100.

Benzene-1,4-dicarboxylic acid (BDC) is the ligand used to build the MIL-101 framework. Highly porous thin films of mesoporous SURMOF-2, including [Cu(bdc)$_2$]$_n$, where bdc = benzene-1,4-dicarboxylic acid, were prepared by liquid-phase epitaxy (Liu et al. 2012; Figure 4.8).

The mesoporous character of the obtained SURMOF-2 porous materials has been demonstrated by small-angle XRD (Liu et al. 2012); while the crystal symmetry remains, the progressively decreasing reflection angles indicate changes in the pore size (Figure 4.9).

In the following study (Hanke et al. 2012), films made of SURMOF-2 [Cu(bdc)$_2$]$_n$ showed remarkable stability in pure water and artificial seawater at 37°C. On the other hand, rapid dissolution of SURMOF-2 occurred in phosphate-buffered saline (PBS), in the PBS-buffered solution of fibrinogen, and in Dulbecco's modified Eagle's medium (DMEM). Those are the typical buffer solutions for cell culture media that are also useful in controlled drug delivery studies. Those buffer solutions contain phosphate ions; therefore, one can conclude that the presence of phosphate ions leads to the chemical destruction of SURMOF-2, due to the leaching of Cu(II) from SURMOF-2 into solution.

Mesoporous UMCM-1 (the University of Michigan Crystalline Material-1) was first synthesized in 2008 (Koh et al. 2008). UMCM-1 has the formula Zn$_4$O(BDC) (BTB)$_{4/3}$, where BDC is the linker of benzenedicarboxylic (terephthalic) acid and BTB is the linker of 1,3,5-tris(4-carboxyphenyl)benzene H$_3$BTB (Figure 4.10). Figure 4.10 in grayscale reproduces Figure 2 in color published in (Koh et al. 2008).

The water vapor adsorption on UMCM-1 was studied at RT and up to 90% relative humidity, and powder XRD and nitrogen adsorption measurements were utilized to check the stability of the sorbent (Schoenecker et al. 2012). UMCM-1 underwent complete loss of crystallinity attributed to the hydrolysis of its four-coordinated zinc–carboxylate framework.

Commercial MOF F300 Basolite is similar to MIL-100(Fe) in its chemical composition and certain structural parameters. Specifically, iron content is 25 versus 21 wt.%, carbon content is 32 versus 29 wt.%, and mesopore size is 21.7 Å versus 25 and 29 Å in F300 and MIL-100(Fe), respectively (Dhakshinamoorthy et al. 2012). The stability of F300 was tested in water at RT as needed for the fabrication of

TABLE 4.7
Stability of Mesoporous MOFs Other Than MIL-101 and MIL-100 in Water, 2012–2014

MOF	Liquid/Vapor?	Liquid Phase	Conditions	Method	Result	Reference
Cu SURMOF $[Cu(BDC)_2]_n$	Liquid	Water, seawater	37°C	XRD, XPS, spectroscopic ellipsometry, IRRAS	Stable	Hanke et al. (2012)
Cu SURMOF $[Cu(BDC)_2]_n$	Liquid	PBS, PBS/fibrinogen, DMEM	37°C	Above	Unstable	Hanke et al. (2012)
UMCM-1 $[Zn_4O(BDC)(BTB)_{4/3}]$	Vapor	—	RT, <90% RH	XRD, N_2 sorption, Water sorption	Unstable	Schoenecker et al. (2012)
F300	Liquid	Water	Ambient	Visual	Stable?	Lee et al. (2014)

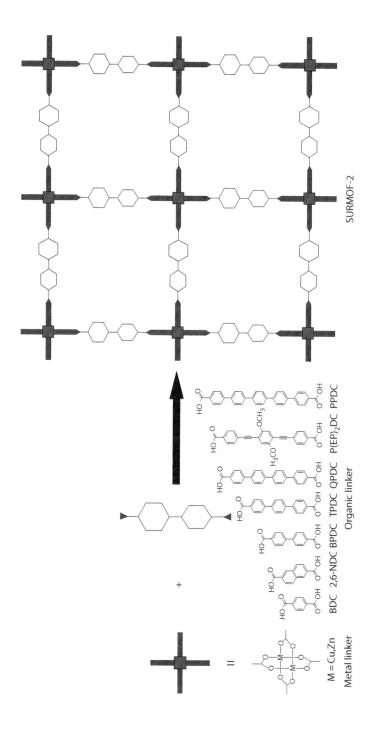

FIGURE 4.8 Schematic representation of the synthesis and formation of SURMOF-2 analogues. (Reprinted by permission from Macmillan Publishers Ltd. *Sci. Rep.*, Liu, J., Lukose, B., Shekhah, O., Arslan, H.K., Weidler, P., Gliemann, H., Brase, S. et al., A novel series of isoreticular metal organic frameworks: Realizing metastable structures by liquid-phase epitaxy, 2, Article number 921. Copyright 2012.)

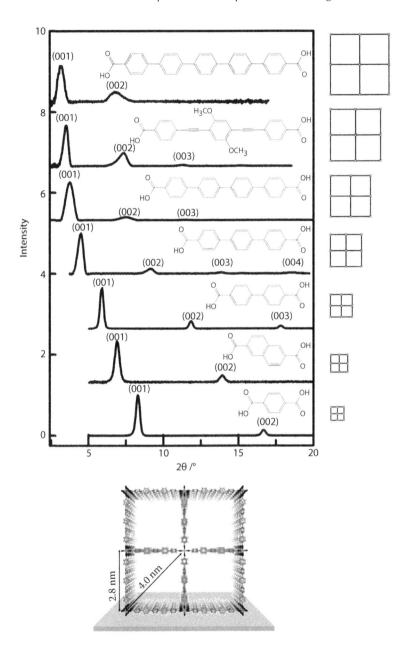

FIGURE 4.9 Out of plane XRD data of Cu-BDC, Cu-2,6-NDC, Cu-BPDC, Cu-TPDC, Cu-QPDC, Cu-P(EP)2DC and Cu-PPDC (upper left), schematic representation (upper right) and proposed structures of SURMOF-2 analogues (lower panel). All the SURMOF-2 are grown on –COOH terminated SAM surface using the LPE method. (Reprinted by permission from Macmillan Publishers Ltd. *Sci. Rep.*, Liu, J., Lukose, B., Shekhah, O., Arslan, H.K., Weidler, P., Gliemann, H., Brase, S. et al., A novel series of isoreticular metal organic frameworks: Realizing metastable structures by liquid-phase epitaxy, 2, Article number 921. Copyright 2012.)

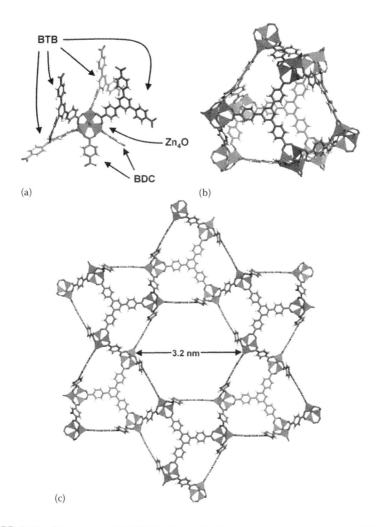

FIGURE 4.10 Structure of UMCM-1: (a) A Zn_4O cluster coordinated to two BDC link-ers and four BTB linkers. Zn_4O clusters, blue tetrahedra; C, gray; H, white; O, red. (b) A microporous cage constructed of six BDC linkers, five BTB linkers, and nine Zn_4O clusters. (c) Structure of UMCM-1 viewed along the C-axis illustrating the one-dimensional mesopore. (From Koh et al. (2008). Copyright Wiley-VCH Verlag GmbH & Co. KGaA. Reproduced with permission.)

porous matrix membranes (Lee et al. 2014). The stability of F300 in water was visu-ally assessed, and F300 "showed relatively high water stability" (Lee et al. 2014). No quantitative measurements of the dissolved versus nondissolved amount of F300 MOF were reported. A more detailed study on the water stability of F300 should be conducted.

In more recent years, the water stability of some newly discovered mesoporous MOFs was accessed (Table 4.8).

TABLE 4.8

Stability of Mesoporous MOFs Other Than MIL-101 and MIL-100 in Water, Papers Since 2014

MOF	Liquid/Vapor?	Liquid Phase	Conditions	Method	Result	Reference
437-MOF	Liquid	Pure water; NaOH in water; HCl in water	25°C–100°C	XRD, N_2 sorption	Pure water: stable; acid or base: unstable	Du et al. (2014)
POST-66(Y)-wt	Liquid	Pure water	25°C–80°C, 24 h	XRD, N_2 sorption, SEM, TEM	Stable	Kim et al. (2015)
PCN-777	Liquid	Pure water; pH = 3; pH = 11		Synchrotron XRD, elemental analysis, FTIR, TEM, TGA	Stable in pure water and at pH = 11	Feng et al. (2015b)
PCN-333(Al)	Liquid	Water at pH = 3–9	RT	Synchrotron XRD, N_2 sorption, Ar sorption, HRTEM	Stable	Feng et al. (2015a)
PCN-333(Fe)	Liquid	Above	Above	Above	Stable	Feng et al. (2015a)

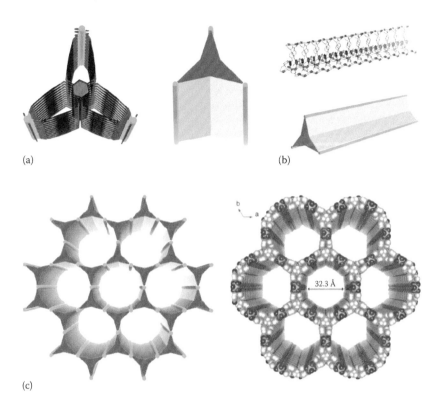

(a)   (b)

(c)

FIGURE 4.11 Crystal structure of 437-MOF. (a) Top and schematic views of the concave triangular-prism supermolecular building block (SBB). (b) Side and schematic views of the concave triangular-prism SBB. (c) View of the 3D framework and the space-filling model showing regular 1D hexagonal channels. (From Du, M., Chen, M., Yang, X.G., Wen, J., Wang, X., Fang, S.M., and Liu, C.S., A channel-type mesoporous In(III)-carboxylate coordination framework with high physicochemical stability for use as an electrode material in supercapacitors, *J. Mater. Chem. A*, 2(25), 9828–9834, 2014. Reproduced by permission of The Royal Society of Chemistry.)

A mesoporous indium-containing 437-MOF with 1D hexagonal channels of ca. 3 nm diameter has been synthesized and characterized (Du et al. 2014; Figure 4.11).

Interestingly, optimal porosity was achieved by heating in boiling water, followed by further heating. The 437-MOF remained stable in liquid water at RT for 30 days and was also found stable in boiling water for >1 day (Du et al. 2014). However, 437-MOF demonstrated a change in the XRD pattern when immersed in aqueous solutions of either NaOH (pH = 9–11) or HCl (pH = 1–3).

There is a significant number of microporous MOFs that are water-stable, while the availability of mesoporous MOFs is very limited. A simple approach for conversion of microporous MOF to water-stable mesoporous MOF would be very desirable. The microporous MOF POST-66(Y) (Figure 4.12) was transformed to the

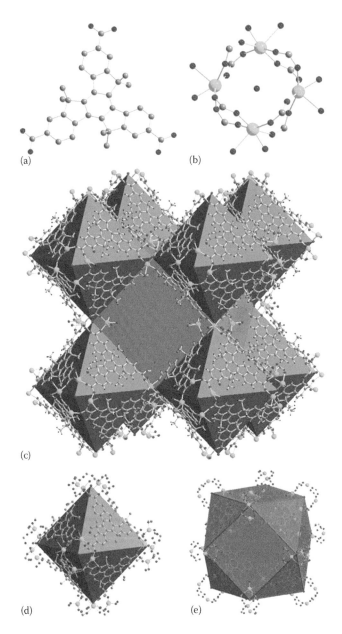

FIGURE 4.12 (a) Structure of hmtt, (b) SBU of POST-66, (c) crystal structure of POST-66, (d) octahedral cage, and (e) cuboctahedral cage. (From Kim, Y., Yang, T., Yun, G., Ghasemian, M.B., Koo, J., Lee, E.Y., Cho, S.J., and Kim, K.: Hydrolytic transformation of microporous metal–organic frameworks to hierarchical micro- and mesoporous MOFs. *Angew. Chem. Int. Ed.* 2015. 54(45). 13273–13278. Copyright Wiley-VCH Verlag GmbH & Co. KGaA. Reproduced with permission.)

hierarchically structured micro- and mesoporous POST-66-wt by simple exposure to liquid water at RT (Kim et al. 2015). The hmtt indicates the linker of methyl-substituted truxene tricarboxylic acid, and the metal node contains Y(III). After immersing POST-66(Y) in water at RT for 24 h, its nitrogen desorption isotherm changed from type I to type IV, which indicates formation of new mesoporous POST-66-wt. The pore size distribution in POST-66-wt as determined by the Barrett–Joyner–Halenda (BJH) method indicated two kinds of mesopores with diameters 3.8 and 13.9 nm.

The dynamics of formation (and stability) of the mesopores were monitored for 24 h by the periodic measurements of nitrogen adsorption/desorption isotherms (Kim et al. 2015). The loss of micropores of POST-66(Y) occurred mostly within 1 h of exposure to liquid water, while mesopores 3–4 nm in diameter were generated during the first 5–10 min. Slow evolution of mesopores occurred within the 24 h period: the number of 3–4 nm mesopores decreased, while the number of >10 nm mesopores increased. Apparently, the obtained mesoporous POST-66-wt is stable in liquid water.

Mesoporous Zr-containing PCN-777 was synthesized (Feng et al. 2015b) start-ing from the heteroaromatic 4,4′,4″-s-triazine-2,4,6-triyl-tribenzoate (TATB) linker (Figure 4.13).

The molecular formula of PCN-777 is $[Zr_6(O)_4(OH)_{10}(H_2O)_6(TATB)_2]$. The molecular size of the TATB linker is about twice the size of the common linker BTC (benzene-1,3,5-tricarboxylate); therefore, the supertetrahedra in PCN-777 are twice as large as those in MIL-100. PCN-777 has the largest mesocage size of 3.8 nm and a pore volume of 2.8 cm^3/g. After being immersed in distilled water and in aqueous solutions with pH = 3 or 11, PCN-777 retained its XRD pattern. The nitrogen adsorption isotherms did not change after treatment in pure water and in the alkaline solution. On the other hand, an exposure to acidic water at pH = 3 caused a decrease in nitrogen adsorption, indicating partial loss of the total surface area (Feng et al. 2015b).

The solvothermal reactions of H_3BTTC (acidic form of benzo-tris-thiophene car-boxylate) or H_3TATB with metal chlorides MCl_3 (M = Al, Fe, V, Sc, In) resulted in the corresponding mesoporous MOFs PCN-332(M) (M = Al, Fe, Sc, V, In) and PCN-333(M) (M = Al, Fe, Sc), respectively (Feng et al. 2015a; Figure 4.14).

The isostructural MOFs PCN-333 have the largest mesocage at 5.5 nm and one of the highest pore volumes at 3.84 cm^3/g amongst all MOFs as reported by 2015. Despite their high porosity, PCN-333(Al) and PCN-333(Fe) remained intact after immersion in liquid water and into the aqueous solutions with pH in the range of 3–9, as was shown by XRD and the nitrogen adsorption isotherms (Feng et al. 2015a). It has been speculated that the high water stability was due to high positive charge den-sity on the rather small cations Al(III) and Fe(III), thus leading to strong electrostatic interactions between metal nodes and organic linkers.

Starting from 2015, there has been an increase in the number of papers on the new kinds of water-stable mesoporous MOFs beyond MIL-100 and MIL-101 families, and many interesting reports are expected in the next few years.

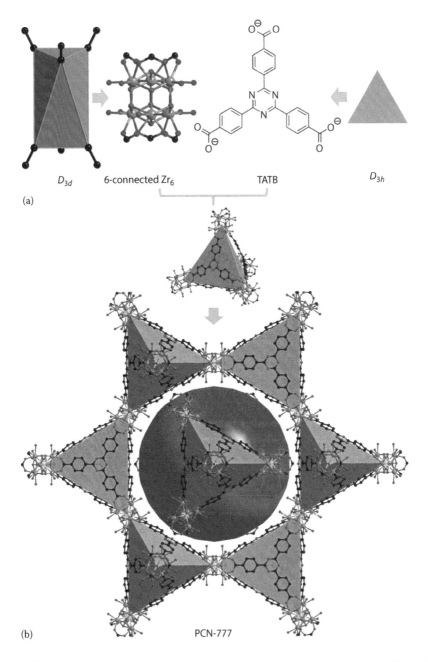

FIGURE 4.13 (a) Six-connected D3d-symmetric Zr_6 antiprismatic units and trigonal–planar organic linker TATB. (b) The structure of PCN-777 with the large mesoporous cage (given in red). (From Feng et al. (2015b). Copyright Wiley-VCH Verlag GmbH & Co. KGaA. Reproduced with permission.)

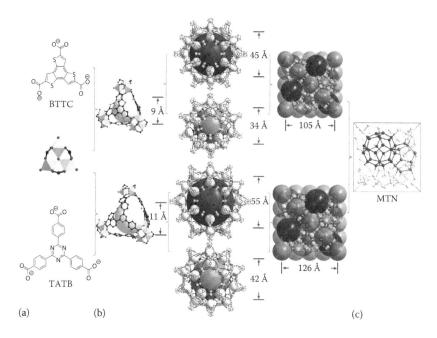

FIGURE 4.14 Structure illustrations of PCN-332 and PCN-333. (a) Ligands used in PCN-332 and PCN-333. (b) Three different cages in PCN-332 and PCN-333. (c) Simplification of PCN-332 and PCN-333 into MTN topology. (Adapted by permission from Macmillan Publishers Ltd. *Nat. Commun.*, Feng, D., Liu, T.-F., Su, J., Bosch, M., Wei, Z., Wan, W., Yuan, D., Chen, Y., Wang, X., Wang, K., Lian, X.Z., Gu, Z.-Y., Park, J., Zou, X., and Zhou, H.-C., Stable metal–organic frameworks containing single-molecule traps for enzyme encapsulation, 6, 5979, copyright 2015.)

REFERENCES

Agostoni, V., T. Chalati, P. Horcajada, H. Willaime, R. Anand, N. Semiramoth, T. Baati et al. 2013. Towards an improved anti-HIV activity of NRTI via metal-organic frameworks nanoparticles. *Advanced Healthcare Materials* 2(12):1630–1637.

Akiyama, G., R. Matsuda, and S. Kitagawa. 2010. Highly porous and stable coordination polymers as water sorption materials. *Chemistry Letters* 39(4):360–361.

Akiyama, G., R. Matsuda, H. Sato, A. Hori, M. Takata, and S. Kitagawa. 2012. Effect of functional groups in MIL-101 on water sorption behavior. *Microporous and Mesoporous Materials* 157:89–93.

Akiyama, G., R. Matsuda, H. Sato, M. Takata, and S. Kitagawa. 2011. Cellulose hydrolysis by a new porous coordination polymer decorated with sulfonic acid functional groups. *Advanced Materials* 23(29):3294–3297.

Bellido, E., M. Guillevic, T. Hidalgo, M. J. Santander-Ortega, C. Serre, and P. Horcajada. 2014. Understanding the colloidal stability of the mesoporous MIL-100(Fe) nanoparticles in physiological media. *Langmuir* 30(20):5911–5920.

Bezverkhyy, I., G. Weber, and J. P. Bellat. 2016. Degradation of fluoride-free MIL-100(Fe) and MIL-53(Fe) in water: Effect of temperature and pH. *Microporous and Mesoporous Materials* 219:117–124.

Bromberg, L., Y. Diao, H. Wu, S. A. Speakman, and T. A. Hatton. 2012. Chromium(III) terephthalate metal organic framework (MIL-101): HF-free synthesis, structure, polyoxometalate composites, and catalytic properties. *Chemistry of Materials* 24(9): 1664–1675.

Čelič, T. B., M. Mazaj, G. Mali, V. Kaučič, and N. Z. Logar. 2013. Hydrothermal stability as an important characteristics of metal-organic framework materials. *Fifth Serbian–Croatian–Slovenian Symposium on Zeolites*. Mountain Zlatibor in Serbia. Website: http://www.zds.org.rs/con-proc.php

Chen, C., M. Zhang, Q. X. Guan, and W. Li. 2012. Kinetic and thermodynamic studies on the adsorption of xylenol orange onto MIL-101(Cr). *Chemical Engineering Journal* 183:60–67.

Cychosz, K. A. and A. J. Matzger. 2010. Water stability of microporous coordination polymers and the adsorption of pharmaceuticals from water. *Langmuir* 26(22):17198–17202.

De Lange, M. F., J. J. Gutierrez-Sevillano, S. Hamad, T. J. H. Vlugt, S. Calero, J. Gascon, and F. Kapteijn. 2013. Understanding adsorption of highly polar vapors on mesoporous MIL-100(Cr) and MIL-101(Cr): Experiments and molecular simulations. *Journal of Physical Chemistry C* 117(15):7613–7622.

Dhakshinamoorthy, A., M. Alvaro, P. Horcajada, E. Gibson, M. Vishnuvarthan, A. Vimont, J. M. Greneche, C. Serre, M. Daturi, and H. Garcia. 2012. Comparison of porous iron trimesates Basolite F300 and MIL-100(Fe) as heterogeneous catalysts for Lewis acid and oxidation reactions: Roles of structural defects and stability. *ACS Catalysis* 2(10):2060–2065.

Du, M., M. Chen, X. G. Yang, J. Wen, X. Wang, S. M. Fang, and C. S. Liu. 2014. A channel-type mesoporous In(III)-carboxylate coordination framework with high physicochemical stability for use as an electrode material in supercapacitors. *Journal of Materials Chemistry A* 2(25):9828–9834.

Ehrenmann, J., S. K. Henninger, and C. Janiak. 2011. Water adsorption characteristics of MIL-101 for heat-transformation applications of MOFs. *European Journal of Inorganic Chemistry* 2011(4):471–474.

Feng, D., T.-F. Liu, J. Su, M. Bosch, Z. Wei, W. Wan, D. Yuan et al. 2015a. Stable metal-organic frameworks containing single-molecule traps for enzyme encapsulation. *Nature Communications* 6:5979.

Feng, D., K. Wang, J. Su, T. F. Liu, J. Park, Z. Wei, M. Bosch, A. Yakovenko, X. Zou, and H. C. Zhou. 2015b. A highly stable zeotype mesoporous zirconium metal-organic framework with ultralarge pores. *Angewandte Chemie—International Edition* 54(1):149–154.

Férey, G., C. Serre, C. Mellot-Draznieks, F. Millange, S. Surblé, J. Dutour, and I. Margiolaki. 2004. A hybrid solid with giant pores prepared by a combination of targeted chemistry, simulation, and powder diffraction. *Angewandte Chemie—International Edition* 43(46):6296–6301.

Férey, G., C. Mellot-Draznieks, C. Serre, F. Millange, J. Dutour, S. Surblé, and I. Margiolaki. 2005. A chromium terephthalate-based solid with unusually large pore volumes and surface area. *Science* 309(5743):2040–2042.

Hanke, M., H. K. Arslan, S. Bauer, O. Zybaylo, C. Christophis, H. Gliemann, A. Rosenhahn, and C. Wöll. 2012. The biocompatibility of metal–organic framework coatings: An investigation on the stability of SURMOFs with regard to water and selected cell culture media. *Langmuir* 28(17):6877–6884.

Hansch, C., A. Leo, S. H. Unger, Ki H. Kim, D. Nikaitani, and E. J. Lien. 1973. Aromatic substituent constants for structure-activity correlations. *Journal of Medicinal Chemistry* 16(11):1207–1216.

Haque, E., V. Lo, A. I. Minett, A. T. Harris, and T. L. Church. 2014. Dichotomous adsorption behaviour of dyes on an amino-functionalised metal-organic framework, amino-MIL-101(Al). *Journal of Materials Chemistry A* 2(1):193–203.

Hong, D. Y., Y. K. Hwang, C. Serre, G. Ferey, and J. S. Chang. 2009. Porous chromium terephthalate MIL-101 with coordinatively unsaturated sites: Surface functionalization, encapsulation, sorption and catalysis. *Advanced Functional Materials* 19(10):1537–1552.

Horcajada, P., R. Gref, T. Baati, P. K. Allan, G. Maurin, P. Couvreur, G. Ferey, R. E. Morris, and C. Serre. 2012. Metal-organic frameworks in biomedicine. *Chemical Reviews* 112(2):1232–1268.

Jeremias, F., A. Khutia, S. K. Henninger, and C. Janiak. 2012. MIL-100(Al, Fe) as water adsorbents for heat transformation purposes—A promising application. *Journal of Materials Chemistry* 22(20):10148–10151.

Khan, N. A., Z. Hasan, and S. H. Jhung. 2013. Adsorptive removal of hazardous materials using metal-organic frameworks (MOFs): A review. *Journal of Hazardous Materials* 244:444–456.

Khutia, A., H. U. Rammelberg, T. Schmidt, S. K. Henninger, and C. Janiak. 2013. Water sorption cycle measurements on functionalized MIL-101Cr for heat transformation application. *Chemistry of Materials* 25(5):790–798.

Kim, Y., T. Yang, G. Yun, M. B. Ghasemian, J. Koo, E. Y. Lee, S. J. Cho, and K. Kim. 2015. Hydrolytic transformation of microporous metal–organic frameworks to hierarchical micro- and mesoporous MOFs. *Angewandte Chemie—International Edition* 54(45):13273–13278.

Koh, K., A. G. Wong-Foy, and A. J. Matzger. 2008. A crystalline mesoporous coordination copolymer with high microporosity. *Angewandte Chemie—International Edition* 47(4):677–680.

Kusgens, P., M. Rose, I. Senkovska, H. Frode, A. Henschel, S. Siegle, and S. Kaskel. 2009. Characterization of metal-organic frameworks by water adsorption. *Microporous and Mesoporous Materials* 120(3):325–330.

Lee, J. Y., C. Y. Y. Tang, and F. W. Huo. 2014. Fabrication of porous matrix membrane (PMM) using metal-organic framework as green template for water treatment. *Scientific Reports* 4. Article number 3740.

Liang, Z. J., M. Marshall, C. H. Ng, and A. L. Chaffee. 2013. Comparison of conventional and HF-free-synthesized MIL-101 for CO_2 adsorption separation and their water stabilities. *Energy & Fuels* 27(12):7612–7618.

Liu, J., B. Lukose, O. Shekhah, H. K. Arslan, P. Weidler, H. Gliemann, S. Brase et al. 2012. A novel series of isoreticular metal organic frameworks: Realizing metastable structures by liquid phase epitaxy. *Scientific Reports* 2. Article number 921.

Low, J. J., A. I. Benin, P. Jakubczak, J. F. Abrahamian, S. A. Faheem, and R. R. Willis. 2009. Virtual high throughput screening confirmed experimentally: Porous coordination polymer hydration. *Journal of the American Chemical Society* 131(43):15834–15842.

Marquez, A. G., A. Demessence, A. E. Platero-Prats, D. Heurtaux, P. Horcajada, C. Serre, J. S. Chang et al. 2012. Green microwave synthesis of MIL-100(Al, Cr, Fe) nanoparticles for thin-film elaboration. *European Journal of Inorganic Chemistry* 2012(32):5165–5174.

Mitchell, L., B. Gonzalez-Santiago, J. P. S. Mowat, M. E. Gunn, P. Williamson, N. Acerbi, M. L. Clarke, and P. A. Wright. 2013. Remarkable Lewis acid catalytic performance of the scandium trimesate metal organic framework MIL-100(Sc) for C–C and C=N bond-forming reactions. *Catalysis Science & Technology* 3(3):606–617.

Nguyen, J. G. and S. M. Cohen. 2010. Moisture-resistant and superhydrophobic metal–organic frameworks obtained via postsynthetic modification. *Journal of the American Chemical Society* 132(13):4560–4561.

Schoenecker, P. M., C. G. Carson, H. Jasuja, C. J. J. Flemming, and K. S. Walton. 2012. Effect of water adsorption on retention of structure and surface area of metal–organic frameworks. *Industrial & Engineering Chemistry Research* 51(18):6513–6519.

Seoane, B., A. Dikhtiarenko, A. Mayoral, C. Tellez, J. Coronas, F. Kapteijn, and J. Gascon. 2015. Metal organic framework synthesis in the presence of surfactants: Towards hierarchical MOFs? *CrystEngComm* 17(7):1693–1700.

Stavitski, E., M. Goesten, J. Juan-Alcañiz, A. Martinez-Joaristi, P. Serra-Crespo, A. V. Petukhov, J. Gascon, and F. Kapteijn. 2011. Kinetic control of metal–organic framework crystallization investigated by time-resolved in situ x-ray scattering. *Angewandte Chemie—International Edition* 50(41):9624–9628.

Taylor-Pashow, K. M. L., J. D. Rocca, Z. Xie, S. Tran, and W. Lin. 2009. Postsynthetic modifications of iron-carboxylate nanoscale metal–organic frameworks for imaging and drug delivery. *Journal of the American Chemical Society* 131(40):14261–14263.

Volkringer, C., D. Popov, T. Loiseau, G. Férey, M. Burghammer, C. Riekel, M. Haouas, and F. Taulelle. 2009. Synthesis, single-crystal x-ray microdiffraction, and NMR characterizations of the giant pore metal-organic framework aluminum trimesate MIL-100. *Chemistry of Materials* 21(24):5695–5697.

Wang, S., L. Bromberg, H. Schreuder-Gibson, and T. A. Hatton. 2013. Organophophorous ester degradation by chromium(III) terephthalate metal-organic framework (MIL-101) chelated to *N,N*-dimethylaminopyridine and related aminopyridines. *ACS Applied Materials and Interfaces* 5(4):1269–1278.

Wickenheisser, M., A. Herbst, R. Tannert, B. Milow, and C. Janiak. 2015. Hierarchical MOF-xerogel monolith composites from embedding MIL-100(Fe,Cr) and MIL-101(Cr) in resorcinol-formaldehyde xerogels for water adsorption applications. *Microporous and Mesoporous Materials* 215:143–153.

Wickenheisser, M., F. Jeremias, S. K. Henninger, and C. Janiak. 2013. Grafting of hydrophilic ethylene glycols or ethylenediamine on coordinatively unsaturated metal sites in MIL-100(Cr) for improved water adsorption characteristics. *Inorganica Chimica Acta* 407:145–152.

Wittmann, T., R. Siegel, N. Reimer, W. Milius, N. Stock, and J. Senker. 2015. Enhancing the water stability of Al-MIL-101-NH$_2$ via postsynthetic modification. *Chemistry—A European Journal* 21(1):314–323.

Wu, B., X. Lin, L. Ge, L. Wu, and T. Xu. 2013. A novel route for preparing highly proton conductive membrane materials with metal-organic frameworks. *Chemical Communications* 49(2):143–145.

Yang, J. F., Y. Wang, L. B. Li, Z. M. Zhang, and J. P. Li. 2015. Protection of open-metal V(III) sites and their associated $CO_2/CH_4/N_2/O_2/H_2O$ adsorption properties in mesoporous V-MOFs. *Journal of Colloid and Interface Science* 456:197–205.

5 Adsorption of Organic Dyes by Mesoporous MOFs in Water

The textile industry is a major consumer of organic dyes, and there are more than 10,000 commercially available synthetic organic dyes (Zollinger 1987). During the industrial coloration process, a large percentage of the applied amount of synthetic dye does not bind to the cloth, and about 10%–15% of the dye used is released into the environment with wastewater (Weber and Adams 1995). Synthetic organic dyes have been specifically designed to be chemically stable toward major environmental factors such as oxygen, water, and sunlight. Therefore, it is difficult to remove synthetic organic dyes from wastewater using "conventional" water treatment procedures. In water, synthetic dyes undergo chemical degradation to form highly toxic and carcinogenic products. Significant research efforts have been directed toward the removal of organic dyes from wastewater (Robinson et al. 2001) using environmentally benign photocatalytic oxidation (Lai et al. 2014) and adsorption (Tong et al. 2012).

The first publication on the adsorption of organic dyes from solution on a metal–organic framework (MOF) was published in 2004 in *Nature* (Chae et al. 2004). Adsorption was performed on MOF-177 from saturated solutions of the dyes in dichloromethane (CH_2Cl_2). MOF-177 is a *microporous* MOF with a maximum pore size of 10.8 Å. The extent of diffusion of dye molecules inside the MOF crystal was assessed by visual observation from the change in the color of MOF-177 on its surface versus the "bulk"; the interior of the MOF was revealed by cleaving the MOF crystal (Chae et al. 2004). Astrazon Orange R and Nile red dyes changed the color of the interior of the MOF crystal rather uniformly, which indicates an easy diffusion of the dye molecules into the bulk of the MOF. This observation is consistent with molecular diameters of those dyes that are close to 11 Å, which is the largest dimension inside the micropore of MOF-177. The adsorption capacity of Astrazon Orange R with a rather small molecular size has been >40 wt.% of the mass of MOF-177, or 16 dye molecules per unit cell of this MOF. In contrast, the larger molecules of Reichardt's dye could penetrate only the outer part of the MOF crystal, but not its bulk. Therefore, surface adsorption allowed only one molecule of Reichardt's dye to be adsorbed, on average, per unit cell of MOF-177 (Chae et al. 2004). The first report on adsorption of organic dye from water onto mesoporous MOF (Haque et al. 2010) was published in 2010.

MOFs, including mesoporous MOFs, contain two main structural units: positively charged metal cations and, usually, negatively charged organic linker units. The majority of organic dyes can be classified into two major groups: cationic dyes and anionic dyes. Therefore, one can assume that the electrostatic interactions between electrically charged structural units of the MOF and cations or anions of organic dyes may play an important role in the adsorption of dyes from water. Therefore, it is convenient to discuss the adsorption of cationic and anionic dyes separately.

5.1   ADSORPTION OF CATIONIC DYES ON MIL-101

Table 5.1 shows the available data on the adsorption of cationic dyes from water on MIL-101.

Hierarchically mesostructured MIL-101(Cr) was synthesized by solvothermal synthesis using cationic surfactant cetyltrimethylammonium bromide (CTAB) as the structural template (Huang et al. 2012). Hierarchically mesostructured MIL-101(Cr) has a well-defined trimodal pore size distribution with mesopores and macropores (Figure 5.1). In caption to Figure 5.1, the DFT refers to the density functional theory applied to calculate the pore size distribution from adsorption data.

Hierarchically mesostructured MIL-101(Cr) demonstrated an accelerated kinetics of adsorption of cationic methylene blue (MB) dye (Figure 5.2) from water versus MIL-101(Cr).

Specifically, a nearly complete adsorption of MB dye on the hierarchically mesostructured MIL-101(Cr) was achieved in 110 min, while an adsorption on "standard" MIL-101(Cr) was as slow as <10% in the same time period (Huang et al. 2012). It was proposed that the significantly accelerated kinetics of adsorption of the dye was due to the combination of the hierarchically structured meso- and macropores. However, the mechanism of interaction of the dye molecule with adsorption sites was not determined.

In order to tune the surface charge distribution in the MIL-101 sorbent, polyoxometalates (POMs) (POM = $K_4PW_{11}VO_{40}$, $H_3PW_{12}O_{40}$, $K_4SiW_{12}O_{40}$) were immobilized onto MIL-101(Cr) resulting (Yan et al. 2014) in the new composite materials POM@MIL-101(Cr). POM@MIL-101(Cr) have been tested in the adsorptive removal of cationic dyes MB and Rhodamine B (RhB) from water versus anionic dye methyl orange (MO). Cationic dyes MB and RhB were adsorbed at much higher rates than anionic MO: 98% of MB and 60% for RhB in 5 min versus just 19% MO. Therefore, the negatively charged POMs immobilized on MIL-101(Cr) are believed to have allowed for a selective adsorption of cationic dyes due to ionic interactions with charged adsorbate molecules (Yan et al. 2014).

Amino-functionalized amino-MIL-101(Al) (Figure 5.3, Haque et al. 2014) was studied for the adsorptive removal of cationic MB versus anionic MO dye from water. The shapes of the adsorption isotherms and the application of the van't Hoff equation to the adsorption equilibrium indicated the spontaneous adsorption of MB dye. On the amino-MIL-101(Al), the adsorption capacity was much higher than on MIL-101(Al). This indicates that the electrostatic interactions between the NH_2 groups of amino-MIL-101(Al) and the cations of the MB dye may have contributed to the higher adsorption capacity.

However, after adsorption of the MB dye, amino-MIL-101(Al) underwent loss of its crystallinity as was found by x-ray diffraction (XRD), and about 30% of Al(III) was leached to the solution (Haque et al. 2014). Significant changes in the "spent" amino-MIL-101(Al) sorbent were also detected by x-ray photoelectron spectroscopy (XPS).

TABLE 5.1

Adsorption of Cationic Organic Dyes on MIL-101 from Water

MOF	Details	Adsorbate Dye	Formula of the Dye	Characterization of the Sorbent	Adsorbed Amount/ Interactions	Reference
MIL-101(Cr)(F)	Activated at 423 K in vacuum	Methylene blue, MB	Figure 5.2	XRD, SEM, TEM, N_2 sorption	10% in 100 min at C_0 = 30 ppm	Huang et al. (2012)
"Hierarchical" MIL-101(Cr)(F)	Activated at 423 K in vacuum	MB	Above	Above	100% in 100 min at C_0 = 30 ppm	Huang et al. (2012)
POM@MIL-101(Cr)(F)	POM = $K_4PW_{11}VO_{40}$, $H_3PW_{12}O_{40}$, $K_4SiW_{12}O_{40}$	MB	Above	XRD, FTIR, TGA, elem. analysis	98% in 5 min at C_0 = 10 mg/L; 371 mg/g; ionic	Yan et al. (2014)
POM@MIL-101(Cr)(F)	Above	Rhodamine B, RhB	Figure 5.2	Above	60% in 5 min at C_0 = 10 mg/L; ionic	Yan et al. (2014)
MIL-101(Al)	Activated	MB	Above	XRD, BET, TGA, XPS	195 mg/g at C_0 = 1000 ppm; ionic	Haque et al. (2014)
MIL-101(Al)-NH_2	Activated at 100°C in vacuum	MB	Above	Above	762 mg/g at C_0 = 1000 ppm; ionic; MOF destruction	Haque et al. (2014)

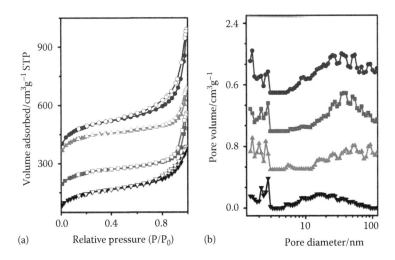

FIGURE 5.1   (a) N_2 adsorption–desorption isotherms at 77 K for the hierarchically meso-structured MIL-101 MOFs C2 (circles), B2 (upward triangles), B1 (squares), and C1 (down-ward triangles); the isotherms for C2 (circles), B2 (upward triangles), and B1 (squares) are vertically offset by 150, 200, and 50 cm³/g, respectively. (b) Distribution of pore diameters obtained using the DFT method; the distributions for C2, B1, and B2 are vertically offset by 1.5, 1.0, and 0.5 cm³/g, respectively. (From Huang et al. (2011). Reproduced by permission of The Royal Society of Chemistry.)

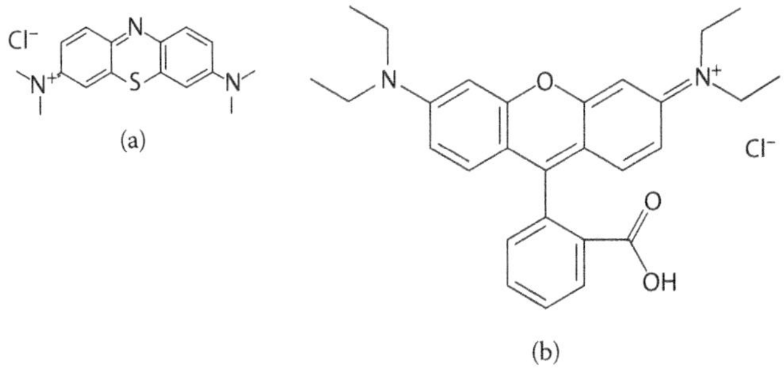

FIGURE 5.2   (a) Methylene blue (MB) and (b) Rhodamine B (RhB) dyes.

Therefore, electrostatic interactions can be considered *too strong* for practical use of adsorption of ionic dyes on amino-MIL-101(Al), since they could cause destruction of the 3D structure of the MOF.

5.2   ADSORPTION OF CATIONIC DYES ON MIL-100

Table 5.2 shows the literature data on the adsorption of cationic organic dyes from water on MIL-100 MOFs.

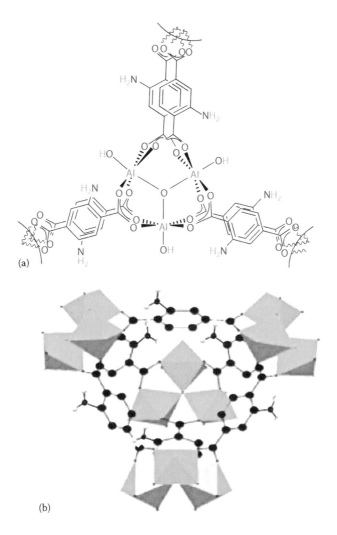

(a)

(b)

FIGURE 5.3 (a) Octahedral cluster of three $AlO_5(OH)$ units with the organic ligand aminoterephthalate. These clusters and ligands form the framework of amino-MIL-101(Al), and each cluster is shown in the bottom diagram as a green octahedron. (b) Supertetrahedral building units of amino-MIL-101(Al) formed by aminoterephthalate ligands and trimeric $AlO_5(OH)$ octahedral clusters. Green, trimeric cluster; red, O; black, C; blue, N; yellow, H. Only the N-bound H atoms are shown. (From Haque et al. (2014). Reproduced by permission of The Royal Society of Chemistry.)

MIL-100(Fe) showed much higher adsorption capacity for cationic dye malachite green (MG) than activated carbon and natural zeolite (Huo and Yan 2012; Figure 5.4).

The kinetics of adsorption obeyed a pseudo-second-order rate law, and the adsorption isotherm of MG followed the Freundlich model. For adsorption of MG on MIL-100(Fe), the enthalpy change was found to be $\Delta H_{ads} > 0$; however, the entropy

TABLE 5.2

Adsorption of Cationic Organic Dyes on MIL-100 from Water

MOF	Activation	Dye	Formula	Characterization	Adsorbed Amount/ Interactions	Reference
MIL-100(Fe)(F)	80°C, vac.	MG	Figure 5.4	XRD, BET, SEM, ζ-potential, XPS	Q_0 = 205 mg/g; Lewis acid	Huo and Yan (2012)
MIL-100(Cr)(F)	120°C	MB	Figure 5.2	XRD, FTIR, DFT, ζ-potential	645.3 mg/g; ionic	Tong et al. (2013)
MIL-100(Fe)(F)	120°C	MB	Above	Above	736.2 mg/g; ionic	Tong et al. (2013)
MIL-100(Fe)(F), made with HF	50°C, vac.	MB	Above	SAXS, XRD, Ar adsorption, SEM, TGA, IR, ζ-potential	1105 mg/g; ionic	Tan et al. (2015)
MIL-100(Fe), HF-free	50°C, vac.	MB	Above	Above	Ionic?	Tan et al. (2015)
MIL-100(Fe) made with H_2SO_4	50°C, vac.	MB	Above	Above	Ionic?	Tan et al. (2015)
MIL-100(Fe)(F)	150°C, air	MB	Above	XRD, BET, TGA, ζ-potential	q_e = 8.0 mg/g at C_0 = 20 mg/L	Jia et al. (2015)

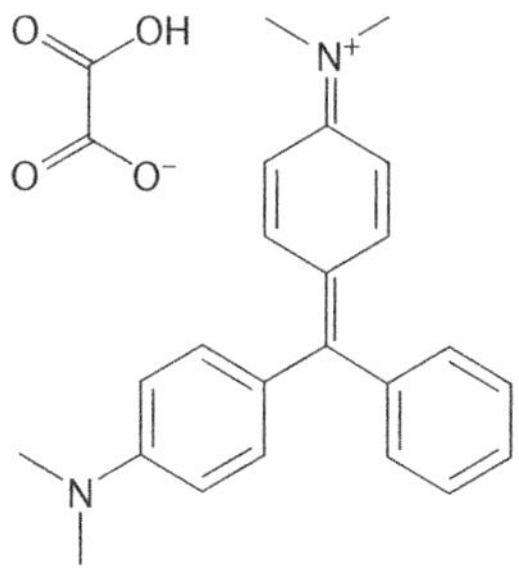

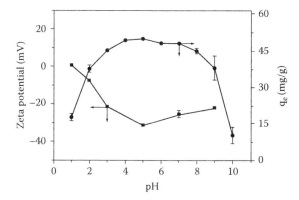

FIGURE 5.4 Malachite green (MG) dye.

FIGURE 5.5 Effect of pH on the adsorption of MG (100 mg/L) on MIL-100(Fe) (10.0 mg) containing 0.01M NaCl, and the zeta potential of MIL-100(Fe) (0.4 mg/L) at 30°C. (From Huo, S.H. and Yan, X.P., Metal-organic framework MIL-100(Fe) for the adsorption of malachite green from aqueous solution, *J. Mater. Chem.*, 22(15), 7449–7455, 2012. Reproduced by permission of The Royal Society of Chemistry.)

change was $\Delta S_{ads} > 0$, which indicates spontaneous adsorption. The comparison of the molecular size of the MG dye and the sizes of microporous windows in MIL-100(Fe) showed that the MG molecule cannot completely enter the mesocavity. To learn about the mechanism, the measurements of the ζ-potential of MIL-100(Fe) in suspension were conducted prior to adsorption at variable pH (Figure 5.5).

Based on the molecular structure, it was speculated that the MG molecule could partially enter the mesocavity through microporous "windows" via $-N(CH_3)_2$ and $-C_6H_5$ functional groups. The measurements of the ζ-potential and XPS data showed that the adsorption of MG dye is driven by electrostatic attraction between the $-N(CH_3)_2$ group in the molecule of the MG dye as the Lewis base, the Fe(III) sites in the MIL-100(Fe), and adsorbed water molecules (Huo and Yan 2012).

Adsorption of cationic MB dye (Figure 5.2) versus anionic MO (Figure 5.6) was compared (Tong et al. 2013) on MIL-100(Cr) and MIL-100(Fe).

As has been reported earlier for MIL-100(Fe) (Huo and Yan 2012) and found by the zeta potential measurements for MIL-100(Cr) (Tong et al. 2013), both MOFs

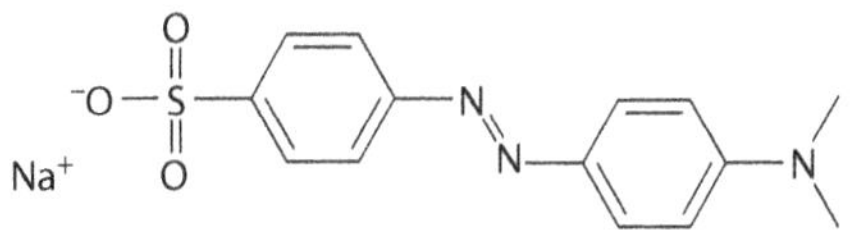

FIGURE 5.6 Methyl orange (MO) dye.

have their surface negatively charged in aqueous solution. Therefore, it was expected that adsorption of cationic dyes would be governed by the strong electrostatic attraction between the cationic dye MB and the negatively charged surfaces of either MIL-100 MOF. MIL-100(Fe) showed significant adsorbed amounts for *both* cationic MB and anionic MO dyes. On the other hand, MIL-100(Cr) *selectively* adsorbed the cationic MB dye from the mixed-water solution of MB and MO. Specifically, MIL-100(Fe) adsorbed 100% MB and 85% MO in water solution, while MIL-100(Cr) adsorbed 100% MB, but only 8% MO under the same conditions (Tong et al. 2013).

MIL-100(Fe) was prepared by three different protocols, namely, by using HF, by using H_2SO_4 instead of HF, and by the fluorine-free method (Tan et al. 2015). These three methods of synthesis yielded several kinds of MIL-100(Fe) with different Langmuir surface areas, total pore volumes, and crystalline sizes. The higher concentration of HF resulted in larger crystals of MIL-100(Fe) within ca. 100 nm to 20 μm, and the adsorption capacity for cationic MB dye increased with the crystalline size (Tan et al. 2015). The adsorption isotherms were fitted with the Langmuir model. The maximum adsorption capacity was as high as 1105 mg/g, which was achieved for MIL-100(Fe)(F) synthesized using the highest content of hydrofluoric acid. The high adsorption capacity has been attributed to the high surface area and suitable pore structure of the MOF sorbent. The adsorption kinetics was fitted with the pseudo-second-order rate law. The kinetic rate of adsorption was found to depend on the particle size of the MOF: the smaller the size, the higher the adsorption rate.

Apart from the particle size, the surface change of the MOF sorbent was found to be important in the adsorption process. The adsorption capacity of MIL-100(Fe)(F) increases rapidly as the pH increases in the range 2–4, then rises more slowly in the pH range 4–8, and finally reaches a plateau at pH = 8–10. In contrast, the ζ-potential of the MIL-100(Fe)(F) sorbent becomes more negative as the pH of the solution increases, indicating that the surface of MOF is negatively charged. Given that MB is a cationic dye, it was concluded that electrostatic interactions play a major role in adsorption. In addition, MIL-100(Fe)(F) was found to be stable at the relatively high pH = 10. The "spent" MIL-100(Fe) sorbent was regenerated by washing with ethanol, and then it was characterized by Fourier transform infrared (FTIR) spectroscopy and XRD, and successfully reused in the adsorption tests. The adsorbed amount of the dye was 403 mg/g after the three cycles, compared to the adsorbed amount at 489 mg/g for the fresh sorbent (Tan et al. 2015).

MIL-100(Fe)(F) was tested in the adsorption of cationic dye MB (Jia et al. 2015). For the MB dye, the relative amounts adsorbed from ultrapure water, rainwater, and river water were 100%, 98.4%, and 100%, respectively. In ultrapure water, at the initial concentration of MB dye within $C_0 = 20$–60 mg/L, the adsorbed amount was $q_e = 8.0$ mg/g at $C_0 = 20$ mg/L and $q_e = 22.9$ mg/g at $C_0 = 60$ mg/L. Adsorption followed a pseudo-second-order kinetics rate law, and it corresponds to the Freundlich model. In contrast to other similar studies with MIL-100(Fe)(F) (e.g., Tan et al. 2015), no significant effects of the pH and ionic strength of the solution were observed for adsorption of the MB dye on MIL-100(Fe)(F) (Jia et al. 2015). The "spent" MIL-100(Fe)(F) sorbent was regenerated and reused once without a significant loss of adsorption capacity.

5.3   ADSORPTION OF ANIONIC DYES ON MIL-101

Published data on adsorption of anionic dyes on MIL-101 MOFs since 2010 are shown in Tables 5.3 and 5.4.

MIL-101(Cr) was studied for adsorptive removal of anionic dye MO from water (Haque et al. 2010). In addition, ethylenediamine-grafted ED–MIL-101(Cr) was obtained by the surface chemical reaction of ethylenediamine with MIL-101(Cr). Further, protonated PED–MIL-101(Cr) was prepared by acidification of ED–MIL-101(Cr) with 0.1 M solution of HCl at room temperature for 6 h. All three MOF sorbents were activated by drying overnight in vacuum at 100°C and were kept in a closed dessicator prior to the adsorption tests. The kinetics of adsorption of the anionic MO dye followed the pseudo-second- or the pseudo-first-order rate law. The adsorptive performance of MIL-101(Cr) improved upon surface modification (Haque et al. 2010). Specifically, the adsorption capacity and kinetic constants of adsorption followed the order MIL-101(Cr) < ED–MIL-101(Cr) < PED–MIL-101(Cr). The adsorption capacity of PED–MIL-101(Cr) decreased with increasing pH of solution of the MO dye in water. This implies that the adsorption of MO is, among other possible factors, due to the electrostatic interactions between the anions of the MO dye and cationic surface groups in the MOF sorbent. However, the type of chemical bonding between the adsorbed dye and structural units(s) in the MOF was not studied by the spectroscopic methods. The application of the van't Hoff equation to the adsorption equilibrium at various temperatures showed that adsorption was a spontaneous endothermic process with $\Delta S_{ads} > 0$. The "spent" sorbents were found to be reusable for adsorption of the MO dye after regeneration by sonication in water (Haque et al. 2010).

Adsorption of anionic dye xylenol orange (XO, Figure 5.7) from water solution on MIL-101(Cr)(F) was studied (Chen et al. 2012). Compared to "common" mesoporous sorbents such as activated carbon and MCM-41, MIL-101(Cr)(F) demonstrated higher adsorption capacity. The adsorption kinetics followed the pseudo-second-order-rate law, and the adsorption thermodynamics followed the Langmuir model.

The adsorbed amount of the dye decreased when the pH of the dye solution was increased (Chen et al. 2012). This observation suggests that ionic interactions might take place between the dye molecule and the structural unit in the MOF. The "spent" MIL-101(Cr)(F) sorbent was regenerated by washing with a dilute solution of NaOH in water. Ionic compound NaOH was apparently chosen for regeneration of the sorbent in order to break the assumed strong ionic bonds between the adsorbed dye molecules and the charged surface of the MOF. However, powder XRD showed that MIL-101(Cr)(F) underwent significant structural changes after the first and second regeneration with the solution of NaOH. Specifically, the characteristic XRD peak

TABLE 5.3
Adsorption of Anionic Dyes on MIL-101 in Water, 2010–2012

MOF	Dye	Formula	Characterization	Adsorbed Amount/Bonding	Reference
MIL-101(Cr)	Methyl orange, MO	Figure 5.6	BET surface area, pore volume	$Q_0 = 114$ mg/g; ionic?	Haque et al. (2010)
ED-MIL-101(Cr)	Methyl orange, MO	Above	Above	$Q_0 = 160$ mg/g; ionic?	Haque et al. (2010)
PED-MIL-101(Cr)	Methyl orange, MO	Above	Above	$Q_0 = 194$ mg/g; ionic?	Haque et al. (2010)
MIL-101(Cr)(F)	Xylenol orange, XO	Figure 5.7	XRD, BET, ζ-potential	$Q_0 = 322$–326 mg/g; ionic? MOF decomposed after regeneration with aqueous NaOH	Chen et al. (2012)

TABLE 5.4
Adsorption of Anionic Dyes on MIL-101 in Water Since 2014

MOF	Dye	Formula of the Dye	Characterization	Adsorbed Amount/Bonding	Reference
MIL-101(Cr)(F)	Uranine	Figure 5.8	SEM, XRD, ζ-potential	126.9 mg/g; ionic interaction by zeta-potential; $\Delta H_{ads} > 0$; $\Delta S_{ads} > 0$	Leng et al. (2014)
Fe_3O_4@MIL-101(Cr), F-free	Congo red (CR)	Figure 5.9	XRD, x-ray microanalysis, N_2 sorption, SEM, TEM	93.8 mg/g at $C_0 = 50$ mg/L	Huang et al. (2014)
Amino–MIL-101(Al)	MO	Figure 5.6	XRD, BET, TGA, XPS	188 mg/g; electrostatic repulsion of –NH_2 in MOF and MO dye?	Haque et al. (2014)

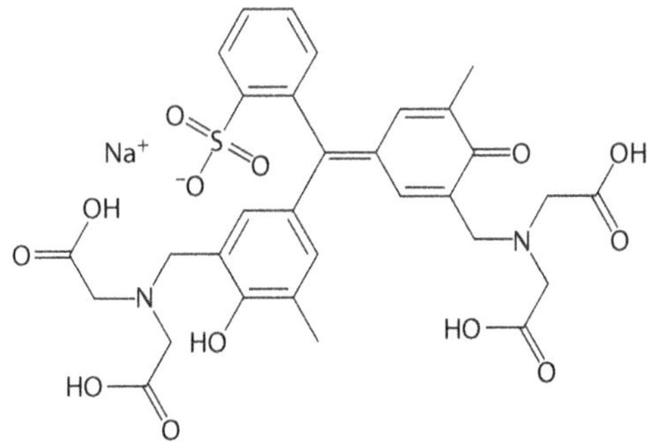

FIGURE 5.7 Xylenol orange (XO) anionic dye.

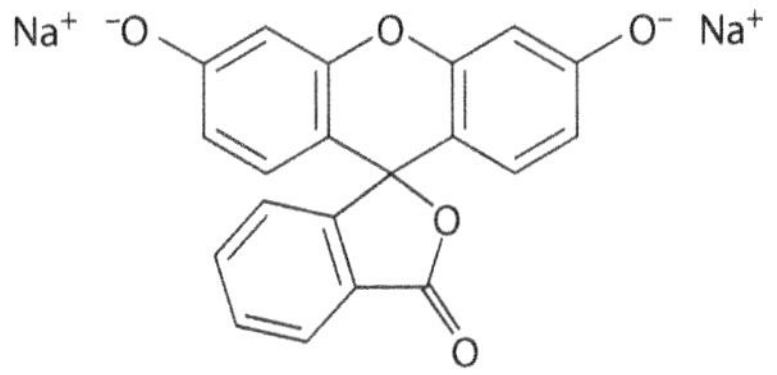

FIGURE 5.8 Uranine dye.

at 2° disappeared, which indicates changes in mesoporosity, while the peak at 8°–10° remained (Chen et al. 2012).

Recent progress in research on adsorption of anionic dyes on MIL-101 can be seen in Table 5.4.

The adsorption of anionic uranine dye (Figure 5.8) from aqueous solution was studied on MIL-101(Cr) (Leng et al. 2014).

The MOF was dried overnight at 150°C in vacuum and was kept in a desiccator before the adsorption tests. The adsorption kinetics obeyed the pseudo-second-order kinetic rate law, and the adsorption isotherm followed the Langmuir model. The adsorption of uranine dye on MIL-101(Cr)(F) was found to be a spontaneous endothermic process, as determined by the van't Hoff equation. With the initial dye concentration varied within 0–400 µmol/L and after the adsorption time of 12 h, the maximum adsorbed amount of 126.9 mg/g MOF was achieved. The zeta potential of MIL-101(Cr)(F) decreased after the adsorption of uranine dye, which suggests electrostatic interactions between MIL-101(Cr)(F) and uranine anions. The "spent" MIL-101(Cr)(F) sorbent was regenerated by sonication in water, and the "fresh" and regenerated MOFs showed no significant difference in the scanning electron microscope (SEM) images. The adsorption capacity of the dye remained about the same after three adsorption/regeneration cycles (Leng et al. 2014).

The nanocomposites Fe_3O_4@MIL-101(Cr) were synthesized by heating an aqueous mixture of $Cr(NO_3)_3$, terephthalic acid, and Fe_3O_4 magnetic nanoparticles without using HF (Huang et al. 2014). The composite sorbent Fe_3O_4@MIL-101 showed the Brunauer–Emmett–Teller (BET) surface area of 1482 m²/g and pore volume of

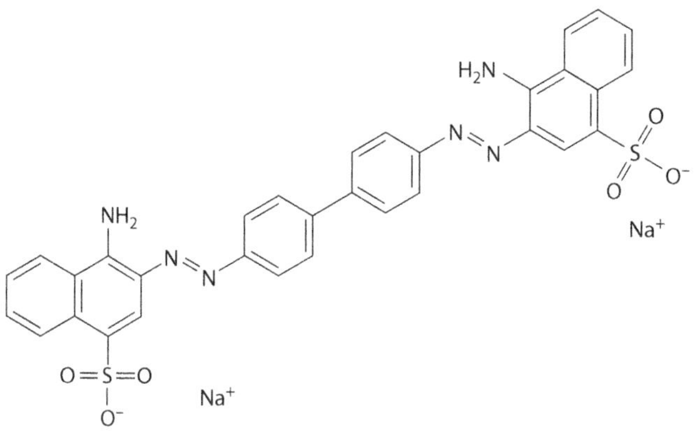

FIGURE 5.9 Congo red dye.

0.822 cm^3/g, which were significantly lower than the respective values for MIL-101(Cr) at 2646 m^2/g and 1.325 cm^3/g, respectively. On the other hand, the incorporation of Fe$_3$O$_4$ did not change the pore size distribution of the mesoporous MIL-101(Cr) host, which is consistent with the size of Fe$_3$O$_4$ nanoparticles at ca. 25 nm, which is much larger than the mesopore sizes in the MOF. Apparently, due to the decrease of the total surface area of the Fe$_3$O$_4$@MIL-101(Cr) nanocomposite versus MIL-101(Cr), the adsorption capacity for the anionic Congo red (CR) dye (Figure 5.9) decreased from 178.6 to 93.8 mg/g at the initial concentration of the dye C$_0$ = 50 mg/L.

The adsorption kinetics was fast, and the adsorption equilibrium was reached in about 20 min (Huang et al. 2014). In the same publication, adsorption on the Fe$_3$O$_4$@ MIL-101(Cr) composite was reported for eight textile dyes other than cationic MB dye and anionic CR, namely, for direct violet, direct dark green, acid orange, acid chrome black T, reactive brilliant blue, reactive yellow, vat pink, and alkali blue. While these dyes were classified as direct dyes, reactive dyes, vat dyes, acid dyes, or cationic dyes, their structural or chemical formulae were not provided. All these dyes were adsorbed on Fe$_3$O$_4$@MIL-101(Cr) within 60 min at the rather low initial concentration of C$_0$ = 50 mg/L (Huang et al. 2014).

Thermally activated amino-MIL-101(Al) was studied for the adsorptive removal of anionic MO dye from aqueous solution (Haque et al. 2014). The adsorption of anionic MO dye on amino-MIL-101(Al) at 188 mg/g sorbent was found to be about four times smaller than that for the cationic MB dye. The lower adsorbed amount of the MO dye was explained by electrostatic repulsion between the anion of MO and the electron-rich –NH$_2$ groups in the 1,4-benzenedicarboxylic acid (BDC) linker of amino-MIL-101(Al). Amino-MIL-101(Al) remained intact after the adsorption of anionic MO dye from water as was found by powder XRD.

5.4 ADSORPTION OF ANIONIC DYES ON MIL-100

Tables 5.5 and 5.6 show published data on adsorption of anionic dyes on MIL-100 and other mesoporous MOFs in water.

TABLE 5.5
Adsorption of Anionic Dyes on MIL-100 and Other Mesoporous MOFs in Water, 2013

MOF	Details	Dye	Formula	Characterization	Adsorbed Amount/Bonding	Reference
MIL-100(Fe)(F)	Activated at 120°C	MO	Figure 5.6	XRD, FTIR, DFT, ζ-potential	1045.2 mg/g; ionic? Adsorbs MO in presence of MB	Tong et al. (2013)
MIL-100(Cr)(F)	Activated at 120°C	MO	Above	Above	211.8 mg/g; ionic? Does not adsorb MO in presence of MB	Tong et al. (2013)
UMCM-1	Not activated	Alizarin red S	Figure 5.10	XRD, ATR-FTIR, SEM	MOF is unstable in water	Halls et al. (2013)

TABLE 5.6
Adsorption of Anionic Dyes on MIL-100 in Water Since 2014

MOF	Details	Dye	Formula	Characterization	Adsorbed Amount/Bonding	Reference
MIL-100(Fe)	F-free, air-dried	Congo red	Figure 5.9	FTIR, STM, TGA/DTG, XRD	714.3 mg/g; ionic?	Moradi et al. (2015)
MIL-100(Fe)(F)	Activated in air at 150°C for 12 h	Methyl blue	Figure 5.11	XRD, BET, BJH, TGA, ζ-potential	$q_e = 3.8$ mg/g at $C_0 = 20$ mg/L	Jia et al. (2015)
Fe_3O_4/MIL-100(Fe)	F-free MIL-100(Fe)	Methyl red	Figure 5.12	XRD, DTA/TG, XPS, TEM, FTIR	$q_e = 586.34$ mg/g at $C_0 = 200$ ppm Ionic interactions, physisorption?	Dadfarnia et al. (2015)

Adsorption of anionic MO versus cationic MB from the water solution was compared on activated MIL-100(Cr)(F) and MIL-100(Fe)(F) (Tong et al. 2013). The adsorption of MO dye followed the pseudo-second-order kinetic rate law and the Langmuir thermodynamic model. For MIL-100(Fe)(F), the adsorption capacity toward the MO dye increased quickly within several minutes and reached equilibrium in 24 h. On the other hand, the adsorption of MO on MIL-100(Cr)(F) was found to be much slower, and the adsorption amount increased steadily in 22 days. To understand the origin of the smaller adsorbed amount of anionic MO dye on MIL-100(Cr)(F), DFT calculations were conducted. The DFT calculations yielded the binding energy of 94.7 kJ/mol for adsorption of water on MIL-100(Cr)(F) and the smaller value of 75.5 kJ/mol on MIL-100(Fe)(F). These data suggest that the competitive adsorption of water is more significant on MIL-100(Cr)(F) than on MIL-100(Fe)(F) (Tong et al. 2013).

The electrochemically active organic dye Alizarin red S (Figure 5.10) was adsorbed (Halls et al. 2013) onto mesoporous MOF UMCM-1 (the University of Michigan Crystalline Material-1) having the molecular formula [$Zn_4O(bdc)(btb)_{4/3}$]. Here, the bdc denotes the 1,4-benzenedicarboxylic acid linker, and btb denotes the 1,3,5-tris(4-carboxyphenyl)benzene linker.

The white powder of UMCM-1 was left in a 3 mM solution of Alizarin red S in water for 72 h. The resultant red powder containing an adsorbed dye was filtered, and its formula was found to be [$Zn_4O(bdc)(btb)_{4/3} \times 0.5$(alizarin red S)]. The attenuated total reflection FTIR (ATR-FTIR) spectra of [$Zn_4O(bdc)(btb)_{4/3} \times 0.5$(alizarin red S)] showed the characteristic peaks of both UMCM-1 and Alizarin red S dye. On the other hand, powder XRD of UMCM-1 without and with adsorbed Alizarin red S dye suggested substantial loss of the long-range order in the lattice of UMCM-1 after absorption of the "guest" molecule Alizarin red S (Halls et al. 2013). One could also note the instability of UMCM-1 in water vapor, as has been reported earlier (Schoenecker et al. 2012). Specifically, adsorption of water vapor on UMCM-1 was studied by powder XRD and nitrogen adsorption measurements at room temperature and up to 90% relative humidity. It was found that UMCM-1 undergoes complete loss of crystallinity in water vapors, presumably due to the hydrolysis of the four-coordinated zinc–carboxylate framework (Schoenecker et al. 2012). Therefore, it is expected that UMCM-1 would undergo significant decomposition in water solution, even in the absence of Alizarin red S.

MIL-100(Fe) was prepared by an F-free method and tested for adsorptive removal of anionic CR dye (Figure 5.9) from water (Moradi et al. 2015). As the pH was increased within the range 4–6, the adsorption capacity increased, then remained

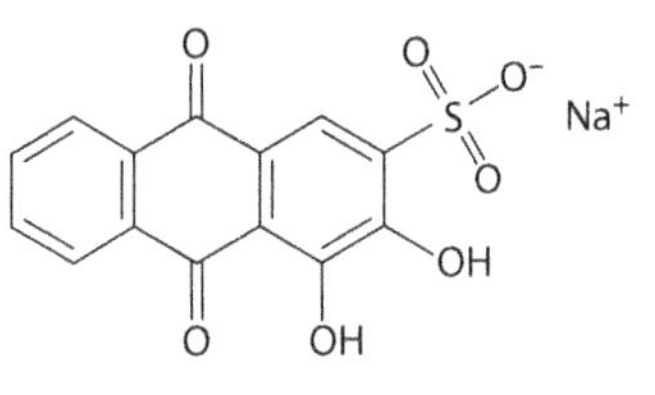

FIGURE 5.10   Alizarin red S dye.

approximately constant within the pH range = 6–8, and finally decreased at pH = 8–10. Two adsorption mechanisms were proposed. At low pH, electrostatic interactions occur between the positively charged surface of the MOF sorbent and the anion of the dye. As the pH increases, a number of negatively charged sites on the surface of MIL-100(Fe) increases, which limits adsorption of this anionic dye. Thus, adsorption at the alkaline pH was assigned to the chemisorption of the CR anions on the Fe(III) sites in MIL-100(Fe). The Elovich equation has been proposed (McLintock 1967) and frequently used (Turner 1975) in the analysis of adsorption in the system "gas/solid," but it has recently been applied in the studies of adsorption in the system "liquid/solid" (Özacar and Şengil 2005), including the adsorption of organic dyes.

The kinetics of adsorption was modeled by the pseudo-first-order rate law, the second-order rate law, the equation for intraparticle diffusion, and the Elovich equation (Moradi et al. 2015). The data of adsorption of the CR dye on MIL-100(Fe) were best fitted with the Elovich formula and with the pseudo-second-order rate law. The adsorption equilibrium was modeled with the Langmuir, Temkin, and Freundlich isotherms; the best results were obtained with the Langmuir isotherm, and the adsorption capacity was as high as 714.3 mg/g sorbent.

MIL-100(Fe)(F) was tested in the adsorption of anionic methyl blue dye (Figure 5.11) from water (Jia et al. 2015).

The adsorption kinetics followed a pseudo-second-order rate law, and adsorption was fitted with the Freundlich isotherm. The changes in the pH and ionic strength did not significantly affect adsorption; the percentages of methyl blue dye adsorbed on MIL-100(Fe)(F) in ultrapure water, rainwater, and river water were 52.1%, 67.5%, and 66.1%, respectively. The adsorption of anionic dye methyl blue proceeded more slowly versus the cationic MB dye, and the adsorbed amounts were also lower. On the other hand, the lower adsorbed amount of anionic methyl blue versus cationic MB dye resulted in the somewhat better reusability of the MIL-100(Fe)(F) sorbent. The adsorption capacity did not significantly decrease after *two* adsorption/desorption cycles for methyl blue dye, compared to just one cycle for the cationic MB dye (Jia et al. 2015).

Functional composite materials Fe_3O_4@MIL-100(Fe) were prepared from nanoparticles of Fe_3O_4 surface-grafted with mercaptoacetic acid and precursors of MIL-100(Fe), namely, $FeCl_3$ and trimethyl 1,3,5-benzenetricarboxylate, without

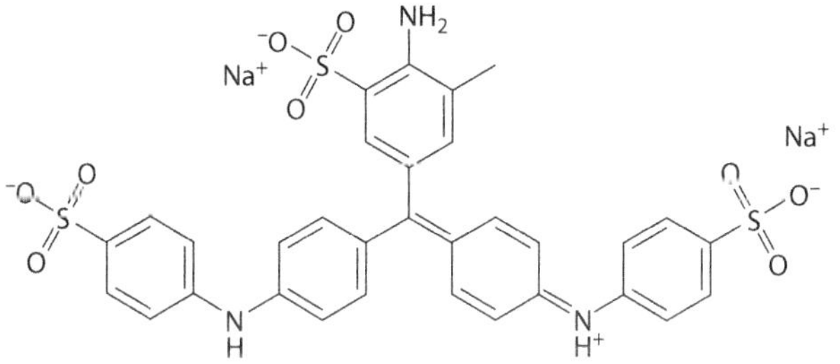

FIGURE 5.11   Structure of methyl blue anionic dye.

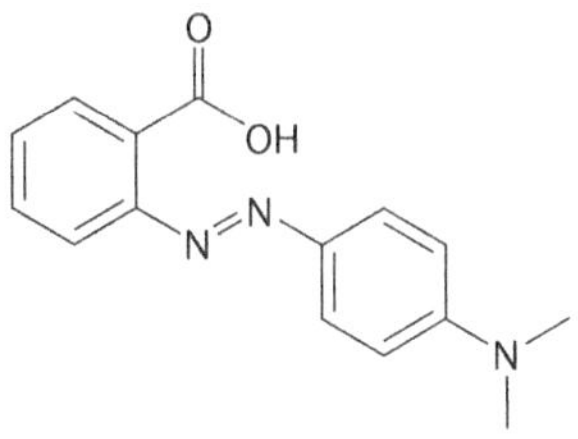

FIGURE 5.12   Methyl red (MR) dye.

using HF (Dadfarnia et al. 2015). The obtained Fe_3O_4@MIL-100(Fe) was studied for adsorptive removal of methyl red (MR) anionic dye (Figure 5.12) from water.

Despite having both the acidic carboxylic and basic amino groups in the molecule, the MR dye is classified as an anionic azo dye (Muthuraman and Teng 2009). MR dyes find extensive use in paper printing, textile dyeing (Lachheb et al. 2002), and as an acid–base indicator. Kinetics of adsorption of the MR dye was found to correspond to the Elovich model and the pseudo-second-order kinetics rate law (Dadfarnia et al. 2015). The amount of adsorbed MR dye was found to be independent of the pH within the range pH = 2–4, and it decreased as the pH of the solution increased up to 10. The decrease in the adsorbed amount was explained by the decrease in the number of positively charged surface sites in the MOF sorbent at higher pH and by repulsion of the MR anion from the surface of the MOF. XPS analysis of Fe_3O_4@MIL-100(Fe) before and after adsorption of MR dye indicated physisorption.

To our knowledge, there is only one paper (Jia et al. 2015) where adsorption of non-ionic vat dye isatin was studied on mesoporous MOF, namely, on MIL-100(Fe) (F). Adsorption was tested with C_0 = 20 mg/L at pH 6.0, and the adsorbed amount of isatin was very small, <10% of that for both cationic MB dye and anionic methyl blue dye. The low adsorbed amount was explained by the lack of electric charge on the molecule of isatin and, hence, the inability of the molecule to adsorb via electrostatic interactions as the major mechanism for (ionic) organic dyes.

Published data on adsorption of cationic and anionic dyes in water solutions on mesoporous MOFs lead to the following conclusions. The major interaction is ionic interaction between the cation or anion of the organic dye and certain structural unit(s) in the MOF with an opposite electric charge. However, the chemical and structural identity of charged adsorption sites in the MOFs often remains unknown. In certain cases, the nature of the interaction between the dye adsorbate and the MOF has been experimentally proven by spectroscopic or zeta-potential measurements. Furthermore, adsorption requires a sterical "match" between the molecular size of the dye and the size of the microporous "windows" in the mesoporous MOFs. The adsorbed amount can be systematically modulated by using polar functional groups as electron donor substituents in the linker of the MOF. However, the obtained stronger interaction between the chemically modified linker and an adsorbed dye molecule may prevent the regeneration of the "spent" MOF sorbent, which would render such a sorbent unusable. It also appears that the research on adsorption of organic dyes has been mostly limited to the MOFs from the MIL-101 and MIL-100 families.

REFERENCES

Chae, H. K., D. Y. Siberio-Perez, J. Kim, Y. Go, M. Eddaoudi, A. J. Matzger, M. O'Keeffe, and O. M. Yaghi. 2004. A route to high surface area, porosity and inclusion of large molecules in crystals. *Nature* 427(6974):523–527.

Chen, C., M. Zhang, Q. X. Guan, and W. Li. 2012. Kinetic and thermodynamic studies on the adsorption of xylenol orange onto MIL-101(Cr). *Chemical Engineering Journal* 183:60–67.

Dadfarnia, S., A. M. H. Shabani, S. E. Moradi, and S. Emami. 2015. Methyl red removal from water by iron based metal-organic frameworks loaded onto iron oxide nanoparticle adsorbent. *Applied Surface Science* 330:85–93.

Halls, J. E., S. D. Ahn, D. M. Jiang, L. L. Keenan, A. D. Burrows, and F. Marken. 2013. Proton uptake vs. redox driven release from metal-organic-frameworks: Alizarin red S reactivity in UMCM-1. *Journal of Electroanalytical Chemistry* 689:168–175.

Haque, E., J. E. Lee, I. T. Jang, Y. K. Hwang, J. S. Chang, J. Jegal, and S. H. Jhung. 2010. Adsorptive removal of methyl orange from aqueous solution with metal-organic frameworks, porous chromium-benzenedicarboxylates. *Journal of Hazardous Materials* 181(1–3):535–542.

Haque, E., V. Lo, A. I. Minett, A. T. Harris, and T. L. Church. 2014. Dichotomous adsorption behaviour of dyes on an amino-functionalised metal-organic framework, amino-MIL-101(Al). *Journal of Materials Chemistry A* 2(1):193–203.

Huang, X. X., L. G. Qiu, W. Zhang, Y. P. Yuan, X. Jiang, A. J. Xie, Y. H. Shen, and J. F. Zhu. 2012. Hierarchically mesostructured MIL-101 metal-organic frameworks: Supramolecular template-directed synthesis and accelerated adsorption kinetics for dye removal. *CrystEngComm* 14(5):1613–1617.

Huang, Y. F., Y. Q. Wang, Q. S. Zhao, Y. Li, and J. M. Zhang. 2014. Facile in situ hydrothermal synthesis of Fe_3O_4@MIL-101 composites for removing textile dyes. *RSC Advances* 4(89):47921–47924.

Huo, S. H. and X. P. Yan. 2012. Metal-organic framework MIL-100(Fe) for the adsorption of malachite green from aqueous solution. *Journal of Materials Chemistry* 22(15):7449–7455.

Jia, Y. Y., Q. Jin, Y. Li, Y. X. Sun, J. Z. Huo, and X. J. Zhao. 2015. Investigation of the adsorption behaviour of different types of dyes on MIL-100(Fe) and their removal from natural water. *Analytical Methods* 7(4):1463–1470.

Lachheb, H., E. Puzenat, A. Houas, M. Ksibi, E. Elaloui, C. Guillard, and J.-M. Herrmann. 2002. Photocatalytic degradation of various types of dyes (Alizarin S, Crocein Orange G, Methyl Red, Congo Red, Methylene Blue) in water by UV-irradiated titania. *Applied Catalysis B: Environmental* 39(1):75–90.

Lai, C. W., J. C. Juan, W. B. Ko, and S. B. A. Hamid. 2014. An overview: Recent development of titanium oxide nanotubes as photocatalyst for dye degradation. *International Journal of Photoenergy* 2014:14. Article ID 524135.

Leng, F., W. Wang, X. J. Zhao, X. Li Hu, and Y. F. Li. 2014. Adsorption interaction between a metal–organic framework of chromium–benzenedicarboxylates and uranine in aqueous solution. *Colloids and Surfaces A: Physicochemical and Engineering Aspects* 441:164–169.

McLintock, I. S. 1967. The Elovich equation in chemisorption kinetics. *Nature* 216(5121):1204–1205.

Moradi, S. E., S. Dadfarnia, A. M. H. Shabani, and S. Emami. 2015. Removal of congo red from aqueous solution by its sorption onto the metal organic framework MIL-100(Fe): Equilibrium, kinetic and thermodynamic studies. *Desalination and Water Treatment* 56(3):709–721.

Muthuraman, G. and T. T. Teng. 2009. Extraction of methyl red from industrial wastewater using xylene as an extractant. *Progress in Natural Science* 19(10):1215–1220.

Özacar, M. and İ. A. Şengil. 2005. A kinetic study of metal complex dye sorption onto pine sawdust. *Process Biochemistry* 40(2):565–572.

Robinson, T., G. McMullan, R. Marchant, and P. Nigam. 2001. Remediation of dyes in textile effluent: A critical review on current treatment technologies with a proposed alternative. *Bioresource Technology* 77(3):247–255.

Schoenecker, P. M., C. G. Carson, H. Jasuja, C. J. J. Flemming, and K. S. Walton. 2012. Effect of water adsorption on retention of structure and surface area of metal–organic frameworks. *Industrial & Engineering Chemistry Research* 51(18):6513–6519.

Tan, F. C., M. Liu, K. Y. Li, Y. R. Wang, J. H. Wang, X. W. Guo, G. L. Zhang, and C. S. Song. 2015. Facile synthesis of size-controlled MIL-100(Fe) with excellent adsorption capacity for methylene blue. *Chemical Engineering Journal* 281:360–367.

Tong, M. M., D. H. Liu, Q. Y. Yang, S. Devautour-Vinot, G. Maurin, and C. L. Zhong. 2013. Influence of framework metal ions on the dye capture behavior of MIL-100(Fe, Cr) MOF type solids. *Journal of Materials Chemistry A* 1(30):8534–8537.

Tong, M. M., X. D. Zhao, L. T. Xie, D. H. Liu, Q. Y. Yang, and C. L. Zhong. 2012. Treatment of waste water using metal-organic frameworks. *Progress in Chemistry* 24(9):1646–1655.

Turner, N. H. 1975. Kinetics of chemisorption: An examination of the Elovich equation. *Journal of Catalysis* 36(3):262–265.

Weber, E. J. and R. L. Adams. 1995. Chemical- and sediment-mediated reduction of the azo dye Disperse Blue 79. *Environmental Science & Technology* 29(5):1163–1170.

Yan, A. X., S. Yao, Y. G. Li, Z. M. Zhang, Y. Lu, W. L. Chen, and E. B. Wang. 2014. Incorporating polyoxometalates into a porous MOF greatly improves its selective adsorption of cationic dyes. *Chemistry—A European Journal* 20(23):6927–6933.

Zollinger, H. 1987. *Colour Chemistry—Synthesis, Properties of Organic Dyes and Pigments.* VCH Publishers, New York.

6 Adsorption of Biologically Active Compounds on Mesoporous MOFs in Water

Trace amounts of medicinal drugs in river waters worldwide (Boyd et al. 2003, Calamari et al. 2003, Tixier et al. 2003, Behera et al. 2011) and in drinking water pose a significant health concern. The removal of medicinal drugs from drinking water is often conducted by adsorption (Ternes et al. 2002). Therefore, the primary aim for the research described in this chapter is to describe the capabilities of mesoporous metal–organic frameworks (MOFs) to effectively remove pharmaceuticals from drinking water at their very low initial concentrations. The second aim is to determine whether mesoporous MOFs can be used for adsorptive removal of small-molecule medicinal drugs and active pharmaceutical ingredients (APIs) from wastewaters of the pharmaceutical industry, where the initial concentrations of the drugs and the APIs are moderate.

In the environmental applications of mesoporous MOFs as sorbents, the major requirements for the choice of MOFs are low toxicity and an affordable price. MOFs with the linkers of 2,4-benzenedicarboxylic (terephthalic) acid, 1,3,5-benzenetricarboxylic (trimesic) acid, and 1,4-naphthalenedicarboxylic acid and with the metal sites formed by Fe, Zn, Ti, Mg, and Ca cations are of relatively low toxicity, due to the rather high LD_{50} values for the respective acids and metal cations. Indeed, nontoxic MOFs have been proposed for use in biomedicine (Horcajada et al. 2010).

MIL-100(Fe) is one of the most well-known mesoporous MOFs with low toxicity (Bellido et al. 2014). Aluminum is a nontoxic metal, and aluminum hydroxide is an active ingredient of many over-the-counter (OTC) antacids. Specifically, an oral LD_{50} of most aluminum salts at 200–1000 mg/kg is comparable to the LD_{50} of calcium at 1000 mg/kg (Farrusseng 2011). Therefore, certain mesoporous MIL-100(Al) (Van de Voorde et al. 2013) and MIL-101(Al) (Haque et al. 2014) are expected to be of low toxicity. Regarding the affordability of mesoporous MOFs as potential industrial-scale sorbents, one can expect a significant price reduction of MIL-101 synthesized from inexpensive terephthalic acid. Terephthalic acid is a large-scale industrial commodity chemical that is produced from the p-xylene fraction from a petroleum refinery. The main industrial use of terephthalic acid is in the production of polyethylene terephthalate (PETE), aka polyester.

The third aim for the reported research is to determine if large amounts of small-molecule drugs, for example, >0.1 g/g sorbent, can readily be adsorbed by mesoporous MOFs. The high adsorbed amounts of biologically active compounds are essential for the controlled encapsulation of medicinal drugs on the MOFs by adsorption (Horcajada et al. 2010, 2012, Keskin and Kizilel 2011, Sun et al. 2013), and the subsequent controlled release of adsorbed (encapsulated) drugs into physiological fluids, cells, or tissues. The major requirement for this emerging application area is a very low toxicity of the MOFs of interest and of molecular products of their metabolism in the human body.

6.1 ADSORPTION OF SMALL-MOLECULE MEDICINAL DRUGS ON MIL-101

Table 6.1 shows published data on the adsorption of small-molecule medicinal drugs on MIL-101 from water solutions.

Naproxen is a nonsteroidal anti-inflammatory drug (NSAID) (Schnitzer et al. 2004) that is available as an OTC medication. The commercial form of naproxen is the sodium salt of (2S)-2-(6-methoxy-2-naphthyl)propanoic acid (Dearmond et al. 1995). The adsorption of naproxen on MIL-101(Cr) from water solution at pH 4.5 has been studied at 25°C (Hasan et al. 2012).

The adsorption capacity of naproxen was found at 132 mg/g for MIL-101(Cr), compared to 81 mg/g for activated carbon. The adsorption kinetics for MIL-101(Cr)

TABLE 6.1

Adsorption of Small-Molecule Medicinal Drugs on Pure and Modified MIL-101 in Aqueous Solutions

MOF	Activated?	Modified?	Adsorbate	Adsorbed Amount	Reference
MIL-101(Cr)(F)	100°C	No	Naproxen	132 mg/g at $C_0 = 10$–15 ppm	Hasan et al. (2012)
MIL-101(Cr)(F)	100°C	No	Clofibric acid	312 mg/g at $C_0 = 50$–150 ppm	Hasan et al. (2012)
MIL-101(Cr)(F)	100°C	No	Naproxen	131 mg/g at $C_0 = 1$–10 ppm	Hasan et al. (2013)
MIL-101(Cr)(F)	Above	No	Clofibric acid	315 mg/g at $C_0 = 5$–50 ppm	Hasan et al. (2013)
MIL-101(Cr)(F)	Above	Acid AMSA	Naproxen	93 mg/g at $C_0 = 1$–10 ppm	Hasan et al. (2013)
AMSA-MIL-101(Cr)(F)	Above	AMSA	Clofibric acid	105 mg/g $C_0 = 5$–50 ppm	Hasan et al. (2013)
ED-MIL-101(Cr)(F)	Above	Base ED	Naproxen	154 mg/g at $C_0 = 1$–10 ppm	Hasan et al. (2013)
ED-MIL-101(Cr)(F)	Above	ED	Clofibric acid	347 mg/g at $C_0 = 5$–50 ppm	Hasan et al. (2013)

followed the pseudo-second-order rate law, and the adsorption isotherm showed the Langmuir shape. The amount of naproxen adsorbed on MIL-101(Cr) decreased when the pH of the naproxen solution was increased in the range pH = 4–12. Given that the acidity constant pK_a of naproxen is ~4, it would mostly exist in the anionic form in this pH range. Therefore, the changes in the adsorbed amount of naproxen versus the pH were not due to the change of the molecular form of naproxen. On the other hand, the density of positive charge on the surface of MIL-101(Cr) decreases when the pH is increased. These considerations resulted in the conclusion that adsorption of naproxen is governed by an ionic interaction between the naproxen anion and unknown positively charged sites in MIL-101(Cr) (Hasan et al. 2012).

Clofibric acid, 2-(4-chlorophenoxy)-2-methylpropanoic acid, is a herbicide (Santos et al. 2000) and a plant growth regulator. Adsorption of clofibric acid from water solution at 25°C was also studied for MIL-101(Cr) (Hasan et al. 2012). MIL-101(Cr) showed the adsorption capacity for clofibric acid at 312 mg/g versus 244 mg/g for activated carbon. The adsorption kinetics followed the pseudo-second-order rate law, and the adsorption isotherm was of the Langmuir shape. As in the case for the adsorption of naproxen, the adsorption of clofibric acid was preferred at lower pH values within the range pH = 3–12. The acidity constant of clofibric acid was not provided by the authors, but it could be found in the literature (Gao and Deshusses 2011); the $pK_a = 2.5$. One can conclude that clofibric acid exists mostly in the anionic form in the whole range pH = 3–12 and thus is likely to be adsorbed as the anion.

In the follow-up study by the same group, the adsorption of naproxen and clofibric acid was studied on MIL-101(Cr) with its surface modified with either acidic or basic functional groups (Hasan et al. 2013). Specifically, MIL-101(Cr) was surface-grafted with aminomethanesulfonic acid (AMSA) or basic ethylenediamine (ED) to prepare the acidic AMSA–MIL-101(Cr) and basic ED–MIL-101(Cr) sorbents, respectively. Before the surface grafting, MIL-101(Cr) was activated at 150°C for 12 h in vacuum to generate the coordinatively unsaturated site (CUS) (Hasan et al. 2013). Then the activated MIL-101(Cr) was allowed to react with ED in anhydrous toluene under reflux for 12 h, filtered, washed with ethanol and deionized (DI) water, and dried at room temperature. The presence of the ED group in the ED–MIL-101(Cr) was confirmed by Fourier transform infrared (FTIR) spectroscopy based on C–N stretching at 1040 cm^{-1}. To prepare the acidic AMSA–MIL-101(Cr), the activated MIL-101(Cr) was refluxed with AMSA in absolute ethanol for 12 h, filtered, washed with DI water, and dried at room temperature. The presence of the AMSA group in AMSA–MIL-101(Cr) was found by FTIR as the S–O stretching bands at 1153 and 1034 cm^{-1}.

The adsorption kinetics follow the pseudo-second-order rate law for naproxene and clofibric acid (Hasan et al. 2013). For naproxen, adsorption capacities were 131 mg/g on MIL-101(Cr), 93 mg/g on acidic AMSA–MIL-101(Cr), and 154 mg/g on basic ED–MIL-101(Cr), compared to 91 mg/g on activated carbon. For clofibric acid, adsorption capacities were significantly higher than for naproxen: 315 mg/g on MIL-101(Cr), 105 mg/g on acidic AMSA–MIL-101(Cr), and 347 mg/g on basic ED–MIL-101(Cr). Obviously, this finding is, at least in part, due to the higher initial concentration of clofibric acid in water (5–50 ppm) compared to naproxen (1–10 ppm). The highest adsorption rate and adsorption capacities were observed on basic ED–MIL-101(Cr). The base-modified ED–MIL-101(Cr) sorbent can be regenerated after adsorption by washing with ethanol, and it can be reused up to three adsorption/ desorption cycles. On the other hand, the adsorbed amount of both drugs on acidic AMSA–MIL-101(Cr) was rather low. The explanation is in repulsion between the anion of naproxen and anionic sites in AMSA–MIL-101(Cr). The adsorption capacity of naproxen on the basic ED–MIL-101(Cr) increased when the pH was increased from 3 to 5, and then decreased at higher pH (Hasan et al. 2013). Such behavior has been explained as being due to the interactions between the acidic form of naproxen (at pH < 4) and the protonated amino groups in ED–MIL-101(Cr). Thus, at the lower pH ~ 3–4, naproxen exists in the acidic (neutral) form, so it does not interact with the protonated amino groups in ED–MIL-101(Cr). On the other hand, the naproxen molecule becomes negatively charged at pH > 5, and the naproxen anion repels the electron-rich $-NH_2$ groups in ED–MIL-101(Cr), which are not protonated at this pH. Thus, the adsorption mechanism was proposed based on the acid–base interactions between the anion of the naproxen and the positively charged or neutral groups in ED–MIL-101(Cr) (Hasan et al. 2013). The main conclusion is that rather high adsorbed amounts of the small-molecule drugs can be achieved in their ionic forms, when the mesoporous MOFs also contain ionic functional groups with an opposite electric charge.

6.2 ADSORPTION OF SMALL-MOLECULE MEDICINAL DRUGS ON MIL-100

Table 6.2 shows published data on adsorption of small-molecule medicinal drugs on MIL-100 in water.

TABLE 6.2
Adsorption of Small-Molecule Medicinal Drugs on Pure MIL-100 in Aqueous Solutions

MOF	Adsorbate	Formula	Adsorbed Amount/ Interactions	Reference
MIL-100(Cr)(F)	Furosemide	Figure 6.1	11.8 mg/g at $C_0 = 7.50$ µg/mL	Cychosz and Matzger (2010)
MIL-100(Cr)(F)	Sulfasalazine	Figure 6.1	6.2 mg/g at $C_0 = 1.40$ µg/mL	Cychosz and Matzger (2010)
MIL-100(Fe)(F)	Naproxen	Figure 6.2	115 mg/g at $C_0 = 10–15$ ppm; ionic	Hasan et al. (2012)
MIL-100(Fe)	Doxorubicin	Figure 6.3	$K_a = 1.8 \pm 0.1 \times 10^4$ M^{-1}; coordination	Anand et al. (2014)
MIL-100(Fe) MIL-100(Cr) MIL-100(Al)	p-Arsanilic acid	Figure 6.4	$Q_0 = 366$ mg/g; coordination	Jun et al. (2015)
MIL-100(Fe) MIL-100(Cr) MIL-100(Al)	Roxarsone	Figure 6.4	$Q_0 = 387$ mg/g; coordination	Jun et al. (2015)

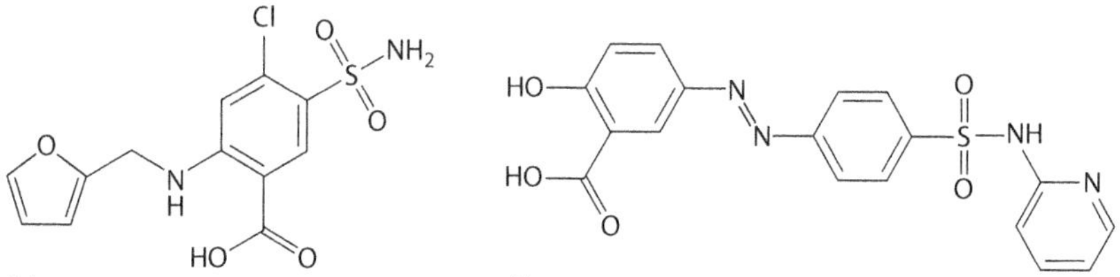

FIGURE 6.1   (a) Furosemide and (b) sulfasalazine.

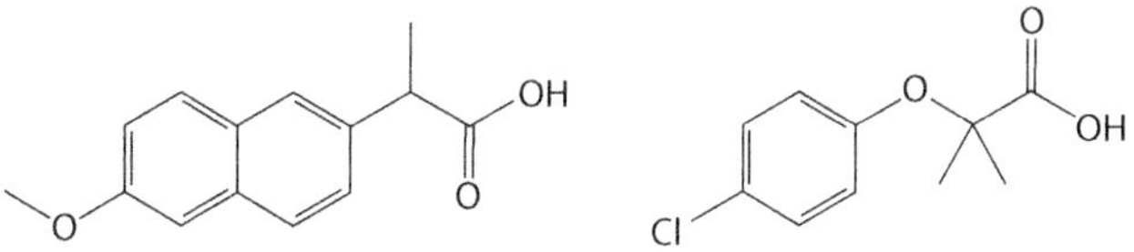

FIGURE 6.2   (a) Naproxen and (b) clofibric acid.

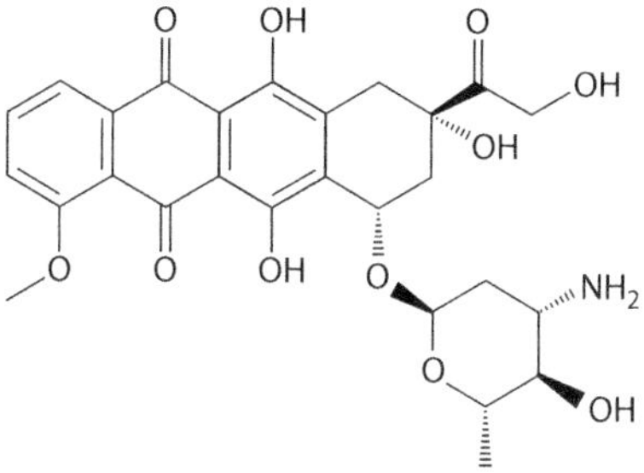

FIGURE 6.3   Doxorubicin.

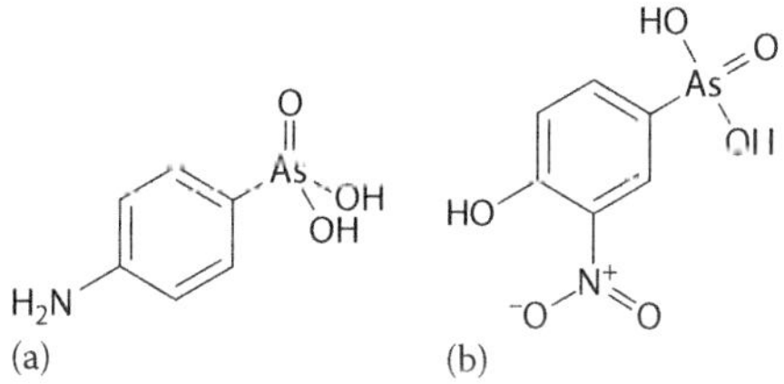

FIGURE 6.4   (a) p-Arsanilic acid (ASA) and (b) roxarsone (ROX).

Furosemide is a diuretic medication (Broekhuysen et al. 1986) that also has some veterinary uses (Hecht et al. 2008).

In prior research, MIL-100(Cr)(F) was shown to be stable in pure liquid water at room temperature for long periods of time (Cychosz and Matzger 2010). For potential use of MOFs to adsorb trace amounts of pharmaceuticals from wastewaters, the adsorption isotherm of furosemide on MIL-100(Cr) was determined at room temperature (Cychosz and Matzger 2010). MIL-100(Cr) showed adsorption capacity at 11.8 mg/g MOF at an initial concentration of the drug 7.50 μg/mL water. After adsorption, powder x-ray diffraction (XRD) showed no change in the structure of MIL-100(Cr) (F). The molecular form of furosemide in the adsorbed state (molecular, anionic, or cationic) has not been determined.

Sulfasalazine (Figure 6.1) is an antirheumatoid drug (Odell et al. 1996). The adsorption capacity of sulfasalazine in water on MIL-100(Cr) was found to be 6.2 mg/g at the initial concentration 1.40 μg/mL; the molecular form of adsorbed sulfasalazine is not known (Cychosz and Matzger 2010). The adsorption of naproxen on MIL-100(Fe) was studied at 25°C at the optimized pH = 4.5 (Hasan et al. 2012). Prior to the adsorption, the as-synthesized MIL-100(Fe) was dried overnight in air; no formation of Fe(III) CUS occurred. The initial concentration of naproxen was varied within 10–15 ppm, and the adsorption capacity was 115 mg/g for MIL-100(Fe), which was higher than that at 81 mg/g for activated carbon. The adsorption kinetics for MIL-100(Fe) followed the pseudo-second-order rate law, and the adsorption isotherms showed the Langmuir shape. The adsorption capacity has been lower on MIL-100(Fe), compared to MIL-101(Cr) (Hasan et al. 2012).

Doxorubicin (DOX, Figure 6.3) is an anticancer drug that has been studied for encapsulation/controlled release by desorption from the nanocrystalline form of the MOF, nanoMIL-100(Fe) (Horcajada et al. 2010).

DOX was adsorbed on nanoMIL-100(Fe) in water or tris(hydroxymethyl)aminomethane, Tris buffer, and the mechanism of interaction of adsorbed DOX with nano-MIL-100(Fe) was investigated (Anand et al. 2014). An adsorption complex was found to have the stoichiometry 1:1 of the DOX:MOF, with an apparent association constant at ca. 1×10^4 M^{-1}. The spectroscopic data indicated that binding occurs via the formation of coordination bonds between one or both deprotonated OH groups of the aglycone moiety in DOX with the CUS Fe(III) centers in MIL-100(Fe). Upon binding, complete quenching of the fluorescence from adsorbed DOX occurred. The data suggested disruption of self-association between the adsorbed DOX molecules.

Roxarsone (ROX) is an organo-arsenic compound that has been widely used as a veterinary drug in growing poultry (Nachman et al. 2013) including chicken farms in the United States. Due to the known carcinogenic effects of both inorganic and organic forms of arsenic compounds and growing consumer concerns, the U.S. Food and Drug Administration (FDA) banned using ROX in production of poultry in 2015.

Adsorption of organo-arsenic compounds p-arsanilic acid (ASA) and ROX with acidic groups $-AsO_3H_2$ has been studied on MIL-100(Fe), MIL-100(Cr), and MIL-100(Al) prepared by "HF-free" synthesis (Jun et al. 2015).

The surface positive charge on these MOFs followed the given trend: MIL-100(Al) > MIL-100(Fe) > MIL-100(Cr). At pH = 4.3, ASA is present as an anion, and the adsorption capacity followed a different trend (Jun et al. 2015):

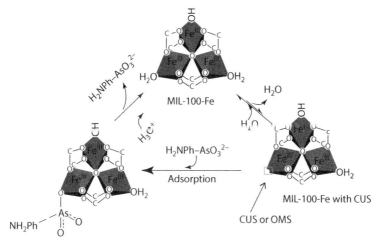

FIGURE 6.5 Suggested mechanism for the adsorption and desorption of ASA over MIL-100-Fe. (Reproduced from Jun, J. W., M. Tong, B. K. Jung, Z. Hasan, C. Zhong, and S. H. Jhung. Effect of central metal ions of analogous metal-organic frameworks on adsorption of organoarsenic compounds from water: Plausible mechanism of adsorption and water purification. *Chem.—A Eur. J.* 2015. 21(1). 347–354 with permission from publisher, Wiley.)

MIL-100(Fe) $\gg$ MIL-100(Al) $\approx$ MIL-100(Cr). Based on this finding, ionic mechanism was ruled out by the authors. Calculations by the density function theory (DFT) predicted adsorption via the arsenate group (Figure 6.5).

6.3 ADSORPTION OF BIOLOGICALLY ACTIVE ORGANIC COMPOUNDS ON MIL-100, MIL-101, AND SIMILAR MOFs

Table 6.3 shows published data on adsorption of miscellaneous large-molecule bio-organic compounds on mesoporous MOFs with 1,4-benzenedicarboxylic acid (BDC) and 1,3,5-benzenetricarboxylic (BTC) linkers.

Creatinine is a small-molecule nitrogen heterocyclic compound that is a uremic toxin often used in diagnostics as a probe molecule to evidence renal failure (Wyss and Kaddurah-Daouk 2000).

The adsorbed amount of creatinine on MIL-100(Fe)(F) at $C_0 = 1$ mM and at 37°C was 190.5 mg/g from phosphate-buffered saline (PBS) (Yang et al. 2014). Such a high adsorption capacity was explained by the suitable size of the microporous windows in MIL-100(Fe) versus the molecular size of creatinine at about $7.1 \times 8.1 \times 3.0$ Å. X-ray photoelectron spectroscopy (XPS) data indicated weak coordination bonds between the nitrogen-containing sites in creatinine (the Lewis base) and Fe(III) sites on MIL-100(Fe) as the Lewis acid. Furthermore, the chemical shift of the C-2 atom (connected to the amino group) in the ^{13}C nuclear magnetic resonance (NMR) spectrum decreased from 169.5 to 168.5 ppm after addition of Fe^{3+}. This suggests that the nitrogen atom in the C-2 position is the site for bonding creatinine to MIL-100(Fe).

TABLE 6.3

Adsorption of Biocompounds on Mesoporous MOFs with BDC and BTC Linkers

MOF	About MOF	Adsorbate	Formula	Characterization	Adsorption/Bonding	Reference
MIL-100(Fe)(F)	Activated in vacuum	Creatinine	Figure 6.6	XRD, TGA, SEM, N_2 ads., XPS, ^1H NMR, ^{13}C NMR	190.5 mg/g; coordination and size match	Yang et al. (2014)
MIL-101-B, $-BO_2H_2$ in linker	Activated at 100°C in vacuum	Galactose, mannose, xylose, glucose	Figure 6.7	XRD, XPS, NMR, TEM, N_2 ads.	Galactose 95 mg/g; cis–diol group with BO_2H_2 in MOF	Zhu et al. (2015)
IPD-mesoMOF-1(Al)-iii	Meso-MIL-100(Al)	Hemoglobin		Elementary analysis, FTIR. XRD, TGA, SEM, N_2 ads.	7 mg/g	Liu et al. (2015)
IPD-mesoMOF-2-ii	Meso-MIL-53(Al)	Hemoglobin		Above	81 mg/g	Liu et al. (2015)
IPD-mesoMOF-2-iii	Meso-MIL-53(Al)	Hemoglobin		Above	94 mg/g	Liu et al. (2015)

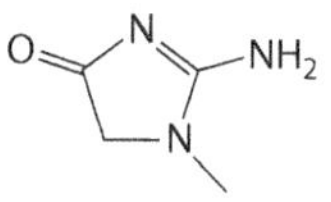

FIGURE 6.6 Structure of model uremic toxin creatinine.

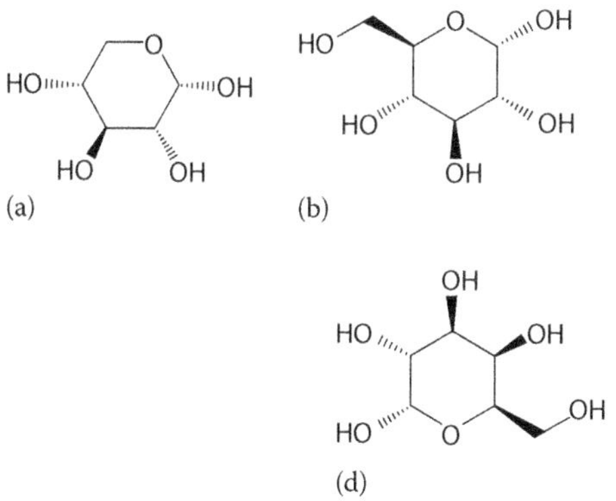

FIGURE 6.7 The C5 and C6 sugars as representative cis–diol-containing biomolecules (CDBs): (a) xylose, (b) glucose, (c) mannose, and (d) galactose.

In the model biological fluid with added human serum albumin (HSA), the adsorbed amount of creatinine decreased due to competitive adsorption on MIL-100. In sorbent regeneration, the desorbed amount of creatinine was as high as 97.6% after three successive 5 min long ultrasonic treatments in methanol. The regenerated MIL-100(Fe) was reused twice with a significant loss in the adsorption capacity (Yang et al. 2014).

Sustainable C6 sugars including glucose (Figure 6.7) are produced on a commercial scale by enzymatic hydrolysis of the cellulose component of plant biomass. Glucose is fermented to bioethanol, which is used as a "green" additive to commercial gasoline. The enzymatic hydrolysis of the hemicellulose component of plant biomass yields sustainable C5 sugars (Fernando et al. 2006) including xylose (Figure 6.7).

Xylose is a feedstock for production of furfural, which is further converted to furfuryl alcohol. The latter is used in the industrial-scale production of sustainable furan resins (Kandola et al. 2015). The separation and purification of the C5 and C6 sugars in aqueous solutions is of significant interest. Based upon the known interactions of boronic acids with sugars (Lü et al. 2013), the boronic acid residue $-BO_2H_2$ present in MOFs was assumed to function as a recognition unit for the adsorption of the cis–diol moieties in CDBs (Zhu et al. 2015). To synthesize boronic group–containing MIL-100-B, metallic Cr, 5-boronobenzene-1,3-dicarboxylic acid (BBDC), BTC, HF, and water were allowed to react in the autoclave. As representative CDBs, galactose, mannose, xylose, and glucose were chosen. At pH 6, the rather small amounts of CDBs were absorbed on MIL-100-B. The adsorbed amounts increased with the increase of the pH and at pH = 9 the adsorbed amount of galactose with the cis–diol group came to 95 mg/g, that is, 85% of the total amount. In the regeneration experiments under acidic conditions, >95% of galactose was desorbed from MIL-100-B in 0.1 M HNO_3 at 25°C for 5 h. Upon regeneration of the "spent" MIL-100-B, the adsorption capacity decreased from 94.2 to 81.7 mg/g. The structure of MIL-100-B was preserved after three adsorption/desorption cycles as judged by XRD analysis (Zhu et al. 2015).

The series of large-cavity mesoporous MOFs with controlled "interparticle-dominated porosity" (IPD)–mesoMOFs were synthesized from precursors of MIL-100, MIL-53, HKUST-1, DUT-5, DUT-4, and MIL-101 (Liu et al. 2015). Compared to the respective "conventional" MOFs of the same nominal chemical composition, much higher porosity up to 2130 m^2/g and mesopore volume up to 2.59 cm^3/g were achieved. The adsorption of hemoglobin (molecular size 6.5 × 5.5 × 5.0 nm) has been studied on IPD–mesoMOFs with mesopores >6.5 nm from phosphate buffer at pH 7. The MOFs of interest were IPD–mesoMOF-1(Al)-iii, IPD–mesoMOF-2-ii, IPD–mesoMOF-2-iii, and IPD–meso-MOF-4-iii. MOFs with larger mesopore size demonstrate higher adsorbed amounts of hemoglobin (Liu et al. 2015).

6.4 ADSORPTION OF LARGE-MOLECULE BIOLOGICALLY ACTIVE COMPOUNDS ON MISCELLANEOUS MESOPOROUS MOFs IN WATER

In order to achieve effective adsorption of large bioactive molecules, one needs to have mesoporous MOFs with large pore sizes, while the choice of available mesoporous MOFs is rather limited. Table 6.4 shows published data on the adsorption of biologically active large-molecule compounds on miscellaneous mesoporous MOFs beyond MIL-100 and MIL-101.

The development of new methods for synthesis of large-pore mesoporous MOFs offers exciting opportunities for the selective adsorption of many classes of (bio) organic compounds. There are a few techniques for the systematic, hypothesis-driven chemical synthesis of new large-pore mesoporous MOFs. First, the larger building units (organic linkers and/or metal clusters) can be used to synthesize MOFs with larger mesopores (Senkovska and Kaskel 2014, Feng et al. 2015).

The new mesoporous MOF, Tb–mesoMOF, was synthesized (Park et al. 2007) by a solvothermal reaction between the large precursor triazine-1,3,5-tribenzoic acid (H_3TATB) and $Tb(NO_3)_3 \cdot 5H_2O$ in N,N-dimethylacetamide (DMA), methanol, and water at 105°C. The structure of the obtained Tb–mesoMOF is shown in Figure 6.8, where ST means supertetrahedron structural element of the MOF.

Tb–mesoMOF (Park et al. 2007) showed high thermal stability and structural rigidity under vacuum, and guest molecules of ferrocene were included by a sublimation procedure at 100°C.

Studies of Tb–mesoMOF for adsorption in aqueous solutions followed. Microperoxidase-11 (MP-11) is a well-known biocatalyst with peroxidase activity (Mondelli et al. 2000). To prepare an immobilized form of MP-11, freshly synthesized

TABLE 6.4
Adsorption of Large-Molecule Biologically Active Compounds on Miscellaneous Mesoporous MOFs in Water

MOF	Adsorbate, Molecular Size	Adsorbed Amount/Interactions	Reference
Tb–mesoMOF	Microperoxidase-11, MP-11	19.1 μmol/g; hydrophobic interactions	Lykourinou et al. (2011)
IRMOF-74-VII	Myoglobin, globular protein, 21 × 35 × 44 Å	One adsorbate molecule per five units of IRMOF-74-VII	Deng et al. (2012)
IRMOF-74-IX	Green fluorescent protein, GFP, barrel structure, diameter 34 Å × length 45 Å	Sterical match?	Deng et al. (2012)
IPD-mesoMOF-4-iii. MOF type of DUT-5	Hemoglobin, 6.5 × 5.5 × 5.0 nm	94 mg/g; sterical match?	Liu et al. (2015)
H-UiO-66(Zr), dried overnight in vacuum at 120°C	DB 86 dye	ca. 240 mg/g; sterical match?	Huang et al. (2015)
H-UiO-66(Zr), see above	MOP-OH	ca. 360 mg/g; sterical match?	Huang et al. (2015)
H-UiO-66(Zr), see above	BSA	ca. 80 mg/g; sterical match?	Huang et al. (2015)

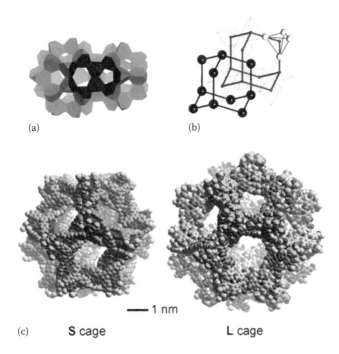

(a)   (b)

—— 1 nm

(c)   **S cage**   **L cage**

FIGURE 6.8 (a) The network of fused S and L mesocages formed by truncated STs. (b) The doubly interpenetrating diamond-like net formed by the L cages (blue spheres) and the centers (small red spheres) of tetrahedra of S cages (yellow spheres). (c) The S and L mesocages drawn as space-filling models. In the S cage, the inner TATB ligands are drawn in red, and outer ligand in blue; in the L cage, the atoms are drawn in their atomic colors. C, gray; H, white; N, blue; O, red; Tb, light blue. (From Park, Y.K., Choi, S.B., Kim, H., Kim, K., Won, B.-H., Choi, K., Choi, J.-S. et al.: Crystal structure and guest uptake of a mesoporous metal–organic framework containing cages of 3.9 and 4.7 nm in diameter. *Angew. Chem. Int. Ed.* 2007. 46(43). 8230–8233. Copyright Wiley-VCH Verlag GmbH & Co. KGaA. Reproduced with permission.)

crystals of Tb–mesoMOF were immersed in a solution of MP-11 in HEPES buffer (4-(2-hydroxyethyl)-1-piperazineethanesulfonic acid in water) and maintained at 37°C (Lykourinou et al. 2011).

The maximum adsorbed amount of 19.1 μmol/g was reached after ~50 h. The obtained adsorption complex MP-11@Tb–mesoMOF was washed with the buffer solution to remove any MP-11 adsorbed on the surface. The UV-Vis diffuse reflectance spectra (DRS) of the adsorption complex has shown the bathochromic (red) shift of the adsorbed MP-11 versus the MP-11 in solution. This is indicative of the interactions between the adsorbed molecules of MP-11 and the hydrophobic interior of the mesocages in Tb–mesoMOF (Lykourinou et al. 2011). Nitrogen adsorption isotherms at 77 K showed a decrease of the total Brunauer–Emmett–Teller (BET) surface area of Tb–mesoMOF from 1935 to 400 m²/g after the adsorption; thus, the majority of the free space in mesopores was occupied by the MP-11 adsorbate. After adsorption, the pore size of MP-11@Tb–mesoMOF was found to be ca. 0.9 nm, in contrast to 3.0–4.1 nm for Tb–mesoMOF before the adsorption. After the adsorption, the catalytic activity of immobilized MP-11 in MP-11@Tb–mesoMOF was assessed via oxidation of 3,5-di-t-butyl-catechol to the o-quinone in water solution. MP-11 did not leach from the adsorption complex MP-11@Tb–mesoMOF for seven catalytic cycles despite the solubility of MP-11 in water. The structural framework of Tb–mesoMOF was maintained as observed by XRD. The stability of the adsorbed MP-11 was attributed to the strong hydrophobic interactions with the Tb–mesoMOF framework (Lykourinou et al. 2011).

Isoreticular metal organic frameworks (IRMOFs) represent a good example of the MOFs designed and made using the "larger linker" principle (Deng et al. 2012). This approach is currently considered a mainstream strategy in the controlled design of the new large-pore mesoporous MOFs (Senkovska and Kaskel 2014). The main disadvantages of this approach are the limits on the unit size in the MOFs at ca. <10 nm (Deng et al. 2012), the lower stability of the MOFs built from the larger precursors of the linkers, interpenetration of the structures leading to a decrease in pore size (Eddaoudi et al. 2002), and the relatively high cost of precursors of the larger linkers.

One of the major synthesis approaches to prepare mesoporous MOFs with large pores is to use controlled "defects" in their structure (Song et al. 2012, Xuan et al. 2012). Precursors of the linkers can easily be available and cheap in this case, but the synthetic methods based on creating defects in the MOFs need to be designed on a case-to-case basis, and these methods can be difficult to reproduce. Yet another, relatively new strategy for the synthesis of mesoporous MOFs with hierarchical pore sizes, potentially including large pores at ca. >3 nm, is using structure-building templates (Qiu et al. 2008). Recently, the synthesis and characterization of new water-stable mesoporous H-UiO-66(Zr) MOFs with pore sizes between 40 and 120 Å were reported (Huang et al. 2015). Specifically, the starting materials were the metal and linker precursors of the microporous MOF-5 and metal and linker precursors of microporous UiO-66(Zr). Here, microporous MOF-5 served as an in situ generated metal–organic assembly (MOA) template. This synthesis strategy is shown in Figure 6.9. MOF-5 was used as the MOA template to be removed by hydrolysis after the synthesis, to yield the final product H-UiO-66(Zr).

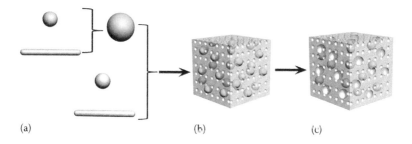

(a)   (b)   (c)

FIGURE 6.9 Schematic representation for the preparation of H-MOF. (a) In situ self-assembly of MOA through the reaction between metal ion and organic ligand. (b) MOA@H-MOF composite formed by one-pot self-assembly reaction. (c) H-MOF formed by removing MOA template. (Adapted from Huang, H. et al., *Nat. Commun.*, 6, Article number 8847, 2015.)

The obtained mesoporous H-UiO-66(Zr) was found to be stable in water as expected for the Zr-containing MOF. The adsorption of the three large-molecule compounds from water was tested as the proof-of-principle experiment (Huang et al. 2015). The adsorbates were organic dye molecule Direct Blue 86 (DB 86) with molecular dimensions ca. $4 \times 12 \times 14$ Å, MOP–OH at ca. $40 \times 40 \times 40$ Å where MOP denotes metal-organic polyhedra molecule, and bovine serum albumin (BSA) protein at ca. $140 \times 40 \times 40$ Å. While the "conventional" microporous UiO-66(Zr) adsorbed up to ca. 40 mg/g for DB 86 dye, the hierarchical H-UiO-66(Zr) with mesopores at 40 Å adsorbed >240 mg/g of this dye. The adsorption of MOP–OH and BSA also had much higher capacities on the new mesoporous H-UiO-66(Zr) versus "conventional" microporous UiO-66(Zr). These new water-stable H-MOFs can find applications in the adsorption/separation of large biomolecules.

In recent decades, there has been growing interest toward the utilization of biocatalysis in the chemical industry. Enzymatic catalysis is one of the major approaches of "green chemistry," and enzymes form one of the most attractive kinds of biocatalysts that can be used in many more competitive technologies compared to "classical" heterogeneous chemical catalysts (Choi et al. 2015). The number of industrial applications of enzymes has increased significantly in recent years, mostly due to advances in protein engineering as well as environmental and purely economic reasons. Therefore, adsorption of large-molecule bio-organic compounds such as enzymes is particularly promising for applications of mesoporous MOFs as structural "hosts," which form new kinds of composite biocatalysts. While mesopores in the MOF sorbents are structurally suitable to retain certain enzyme molecules, one of the "bottlenecks" in the wider research and applications of mesoporous MOFs is that the microporous "windows" in MIL-100 and MIL-101 are too small to allow admission of an enzyme adsorbate to the mesopores. The obvious solution is to investigate the new kinds of mesoporous MOFs beyond these two "benchmark" MOF families. Yet another difficulty in the systematic study of adsorption and desorption of large-molecule adsorbates in mesoporous MOFs is that it is often impossible to determine the specific functional group that would be *mainly* responsible for chemical bonding to the MOF. Similarly, the identity of the pertinent adsorption site(s) in the MOF sorbent can be difficult to reveal. Instead, research can be focused on the changes in the large adsorbate molecule itself upon its sorption within the large mesopore, for example, its molecular or ionic state, the counter ions coadsorbed from the solution, on the conformational changes of the adsorbate molecule, and so on.

REFERENCES

Anand, R., F. Borghi, F. Manoli, I. Manet, V. Agostoni, P. Reschiglian, R. Gref, and S. Monti. 2014. Host-guest interactions in Fe(III)-trimesate MOF nanoparticles loaded with doxorubicin. *Journal of Physical Chemistry B* 118(29):8532–8539.

Behera, S. K., H. W. Kim, J. E. Oh, and H. S. Park. 2011. Occurrence and removal of antibiotics, hormones and several other pharmaceuticals in wastewater treatment plants of the largest industrial city of Korea. *Science of the Total Environment* 409(20):4351–4360.

Bellido, E., M. Guillevic, T. Hidalgo, M. J. Santander-Ortega, C. Serre, and P. Horcajada. 2014. Understanding the colloidal stability of the mesoporous MIL-100(Fe) nanoparticles in physiological media. *Langmuir* 30(20):5911–5920.

Boyd, G. R., H. Reemtsma, D. A. Grimm, and S. Mitra. 2003. Pharmaceuticals and personal care products (PPCPs) in surface and treated waters of Louisiana, USA and Ontario, Canada. *Science of the Total Environment* 311(1–3):135–149.

Broekhuysen, J., F. Deger, J. Douchamps, H. Ducarne, and A. Herchuelz. 1986. Torasemide, a new potent diuretic—Double-blind comparison with furosemide. *European Journal of Clinical Pharmacology* 31:29–34.

Calamari, D., E. Zuccato, S. Castiglioni, R. Bagnati, and R. Fanelli. 2003. Strategic survey of therapeutic drugs in the rivers Po and Lambro in northern Italy. *Environmental Science & Technology* 37(7):1241–1248.

Choi, J.-M., S.-S. Han, and H.-S. Kim. 2015. Industrial applications of enzyme biocatalysis: Current status and future aspects. *Biotechnology Advances* 33(7):1443–1454.

Cychosz, K. A. and A. J. Matzger. 2010. Water stability of microporous coordination polymers and the adsorption of pharmaceuticals from water. *Langmuir* 26(22):17198–17202.

Dearmond, B., C. A. Francisco, J. S. Lin, F. Y. Huang, S. Halladay, R. D. Bartizek, and K. L. Skare. 1995. Safety profile of over-the-counter naproxen sodium. *Clinical Therapeutics* 17(4):587–601.

Deng, H. X., S. Grunder, K. E. Cordova, C. Valente, H. Furukawa, M. Hmadeh, F. Gandara et al. 2012. Large-pore apertures in a series of metal-organic frameworks. *Science* 336(6084):1018–1023.

Eddaoudi, M., J. Kim, N. Rosi, D. Vodak, J. Wachter, M. O'Keeffe, and O. M. Yaghi. 2002. Systematic design of pore size and functionality in isoreticular MOFs and their application in methane storage. *Science* 295(5554):469–472.

Farrusseng, D., ed. 2011. *Metal-Organic Frameworks: Applications from Catalysis to Gas Storage*, Weinheim, Germany: Wiley-VCH.

Feng, D. W., K. C. Wang, J. Su, T. F. Liu, J. Park, Z. W. Wei, M. Bosch, A. Yakovenko, X. D. Zou, and H.-C. Zhou. 2015. A highly stable zeotype mesoporous zirconium metal-organic framework with ultralarge pores. *Angewandte Chemie—International Edition* 54(1):149–154.

Fernando, S., S. Adhikari, C. Chandrapal, and N. Murali. 2006. Biorefineries: Current status, challenges, and future direction. *Energy & Fuels* 20(4):1727–1737.

Gao, Y. H. and M. A. Deshusses. 2011. Adsorption of clofibric acid and ketoprofen onto powdered activated carbon: Effect of natural organic matter. *Environmental Technology* 32(15):1719–1727.

Haque, E., V. Lo, A. I. Minett, A. T. Harris, and T. L. Church. 2014. Dichotomous adsorption behaviour of dyes on an amino-functionalised metal-organic framework, amino-MIL-101(Al). *Journal of Materials Chemistry A* 2(1):193–203.

Hasan, Z., E. J. Choi, and S. H. Jhung. 2013. Adsorption of naproxen and clofibric acid over a metal-organic framework MIL-101 functionalized with acidic and basic groups. *Chemical Engineering Journal* 219:537–544.

Hasan, Z., J. Jeon, and S. H. Jhung. 2012. Adsorptive removal of naproxen and clofibric acid from water using metal-organic frameworks. *Journal of Hazardous Materials* 209:151–157.

Hecht, S., I. F. Lane, G. B. Daniel, F. Morandi, and D. E. Sharp. 2008. Diuretic renal scintigraphy in normal cats. *Veterinary Radiology & Ultrasound* 49(6):589–594.

Horcajada, P., T. Chalati, C. Serre, B. Gillet, C. Sebrie, T. Baati, J. F. Eubank et al. 2010. Porous metal-organic-framework nanoscale carriers as a potential platform for drug delivery and imaging. *Nature Materials* 9(2):172–178.

Horcajada, P., R. Gref, T. Baati, P. K. Allan, G. Maurin, P. Couvreur, G. Ferey, R. E. Morris, and C. Serre. 2012. Metal-organic frameworks in biomedicine. *Chemical Reviews* 112(2):1232–1268.

Huang, H., J.-R. Li, K. Wang, T. Han, M. Tong, L. Li, Y. Xie, Q. Yang, D. Liu, and C. Zhong. 2015. An in situ self-assembly template strategy for the preparation of hierarchical-pore metal-organic frameworks. *Nature Communications* 6:Article number 8847.

Jun, J. W., M. Tong, B. K. Jung, Z. Hasan, C. Zhong, and S. H. Jhung. 2015. Effect of central metal ions of analogous metal-organic frameworks on adsorption of organoarsenic compounds from water: Plausible mechanism of adsorption and water purification. *Chemistry—A European Journal* 21(1):347–354.

Kandola, B., J. Ebdon, and K. Chowdhury. 2015. Flame retardance and physical properties of novel cured blends of unsaturated polyester and furan resins. *Polymers* 7(2):298–315.

Keskin, S. and S. Kizilel. 2011. Biomedical applications of metal organic frameworks. *Industrial & Engineering Chemistry Research* 50(4):1799–1812.

Liu, X., Z. Q. Shen, H. H. Xiong, Y. Chen, X. N. Wang, H. Q. Li, Y. T. Li, K. H. Cui, and Y. Q. Tian. 2015. Hierarchical porous materials based on nanoscale metal-organic frameworks dominated with permanent interparticle porosity. *Microporous and Mesoporous Materials* 204:25–33.

Lü, C., H. Li, H. Wang, and Z. Liu. 2013. Probing the interactions between boronic acids and cis-diol-containing biomolecules by affinity capillary electrophoresis. *Analytical Chemistry* 85(4):2361–2369.

Lykourinou, V., Y. Chen, X.-S. Wang, L. Meng, T. Hoang, L.-J. Ming, R. L. Musselman, and S. Ma. 2011. Immobilization of MP-11 into a mesoporous metal–organic framework, MP-11@mesoMOF: A new platform for enzymatic catalysis. *Journal of the American Chemical Society* 133(27):10382–10385.

Mondelli, R., L. Scaglioni, S. Mazzini, G. Bolis, and G. Ranghino. 2000. 3D structure of microperoxidase-11 by NMR and molecular dynamic studies. *Magnetic Resonance in Chemistry* 38(4):229–240.

Nachman, K. E., P. A. Baron, G. Raber, K. A. Francesconi, A. Navas-Acien, and D. C. Love. 2013. Roxarsone, inorganic arsenic, and other arsenic species in chicken: A U.S.-based market basket sample. *Environmental Health Perspectives* 121(7):818–824.

Odell, J. R., C. E. Haire, N. Erikson, W. Drymalski, W. Palmer, P. J. Eckhoff, V. Garwood et al. 1996. Treatment of rheumatoid arthritis with methotrexate alone, sulfasalazine and hydroxychloroquine, or a combination of all three medications. *New England Journal of Medicine* 334(20):1287–1291.

Park, Y. K., S. B. Choi, H. Kim, K. Kim, B.-H. Won, K. Choi, J.-S. Choi et al. 2007. Crystal structure and guest uptake of a mesoporous metal–organic framework containing cages of 3.9 and 4.7 nm in diameter. *Angewandte Chemie International Edition* 46(43):8230–8233.

Qiu, L.-G., T. Xu, Z.-Q. Li, W. Wang, Y. Wu, X. Jiang, X.-Y. Tian, and L. D. Zhang 2008. Hierarchically micro- and mesoporous metal–organic frameworks with tunable porosity. *Angewandte Chemie International Edition* 47(49):9487–9491.

Santos, T. C. R., J. C. Rocha, and D. Barcelo. 2000. Determination of rice herbicides, their transformation products and clofibric acid using on-line solid-phase extraction followed by liquid chromatography with diode array and atmospheric pressure chemical ionization mass spectrometric detection. *Journal of Chromatography A* 879(1):3–12.

Schnitzer, T. J., G. R. Burmester, E. Mysler, M. C. Hochberg, M. Doherty, E. Ehrsam, X. Gitton et al. 2004. Comparison of lumiracoxib with naproxen and ibuprofen in the Therapeutic Arthritis Research and Gastrointestinal Event Trial (TARGET), reduction in ulcer complications: Randomised controlled trial. *Lancet* 364(9435):665–674.

Senkovska, I. and S. Kaskel. 2014. Ultrahigh porosity in mesoporous MOFs: Promises and limitations. *Chemical Communications* 50(54):7089–7098.

Song, L. F., J. Zhang, L. X. Sun, F. Xu, F. Li, H. Z. Zhang, X. L. Si et al. 2012. Mesoporous metal-organic frameworks: Design and applications. *Energy & Environmental Science* 5(6):7508–7520.

Sun, C. Y., C. Qin, X. L. Wang, and Z. M. Su. 2013. Metal-organic frameworks as potential drug delivery systems. *Expert Opinion on Drug Delivery* 10(1):89–101.

Ternes, T. A., M. Meisenheimer, D. McDowell, F. Sacher, H.-J. Brauch, B. Haist-Gulde, G. Preuss, U. Wilme, and N. Zulei-Seibert. 2002. Removal of pharmaceuticals during drinking water treatment. *Environmental Science & Technology* 36(17):3855–3863.

Tixier, C., H. P. Singer, S. Oellers, and S. R. Muller. 2003. Occurrence and fate of carbamazepine, clofibric acid, diclofenac, ibuprofen, ketoprofen, and naproxen in surface waters. *Environmental Science & Technology* 37(6):1061–1068.

Van de Voorde, B., M. Boulhout, F. Vermoortele, P. Horcajada, D. Cunha, J. S. Lee, J. S. Chang et al. 2013. N/S-heterocyclic contaminant removal from fuels by the mesoporous metal-organic framework MIL-100: The role of the metal ion. *Journal of the American Chemical Society* 135(26):9849–9856.

Wyss, M. and R. Kaddurah-Daouk. 2000. Creatine and creatinine metabolism. *Physiological Reviews* 80(3):1107–1213.

Xuan, W., C. Zhu, Y. Liu, and Y. Cui. 2012. Mesoporous metal-organic framework materials. *Chemical Society Reviews* 41(5):1677–1695.

Yang, C. X., C. Liu, Y. M. Cao, and X. P. Yan. 2014. Metal-organic framework MIL-100(Fe) for artificial kidney application. *RSC Advances* 4(77):40824–40827.

Zhu, X. Y., J. L. Gu, J. Y. Zhu, Y. S. Li, L. M. Zhao, and J. L. Shi. 2015. Metal-organic frameworks with boronic acid suspended and their implication for cis-diol moieties binding. *Advanced Functional Materials* 25(25):3847–3854.

7 Adsorption of Miscellaneous Organic Compounds in Water

Adsorption studies of miscellaneous organic compounds from water are conducted with the aim of purification of water from spilled liquid fossil fuels and various by-products of industrial organic chemical synthesis. These studies on mesoporous metal–organic frameworks (MOFs) are rather limited. Adsorption of benzene on MIL-101(Cr) was tested at 25°C in an aqueous solution at the initial concentration $C_0 = 1000$ ppm (Jhung et al. 2007); see also Table 7.1.

MIL-101(Cr) adsorbed up to 0.8 g benzene/g sorbent, which was approximately two times higher than the adsorption capacity of activated carbon (AC) under the same conditions. Adsorption of benzene on MIL-101(Cr) also occurred faster than on AC. This effect could be due to the larger pore diameter or the smaller particle size, as has been mentioned by the authors; however, the exact cause of this difference has not been determined in this early study (Jhung et al. 2007).

The three aromatic compounds with low acidity or basicity, namely, thiophene (Th), pyrrole (Py), and nitrobenzene (NB), were adsorbed from water on MIL-101(Cr) and, for comparison, on AC (Bhadra et al. 2015). The maximum adsorbed amounts followed the order NB > Py > Th. On AC, this order was different: Th > NB > Py. To learn more about the effect of the solvent, the adsorbed amount was studied versus the dielectric constant of the solvent, which was water, n-butanol, and n-octane. It was found that AC preferentially adsorbed three aromatic compounds from water (the highly polar solvent), while MIL-101(Cr) was very effective in adsorbing from the nonpolar n-octane (Figure 7.1).

The adsorption capacity of all three aromatic compounds studied on MIL-101(Cr) in water was found to be rather low due to the competitive adsorption of polar water, as was confirmed by measuring the hydrophobicity indexes via the adsorption experiments in the water–toluene vapor mixtures (Bhadra et al. 2015).

In order to increase the adsorption capacity or selectivity of adsorption in solution, postsynthetic modification (PSM) of mesoporous MOFs was employed (Table 7.2).

Several anion-exchanged MIL-101(Cr) were synthesized by PSM of fluorine-free MIL-101(Cr); alternatively, the synthesis was started with the modified linker using the so-called preassembled modification (PAM) method (Liu et al. 2015). Specifically, the new mesoporous MOF MIL-101-DMEN was prepared from hydrated MIL-101 as given for the published procedure (Hong et al. 2009), via the reaction of N,N-dimethylethylenediamine (DMEN) in anhydrous toluene, washed, and dried in vacuum at 100°C. The amino group of DMEN was believed to have reacted with the Cr(III) coordinatively unsaturated site (CUS). Next, its quaternized

TABLE 7.1
Adsorption of Miscellaneous Organic Compounds on Pure MIL-101(Cr) in Aqueous Solutions

MOF	Details of MOF	Adsorbate	Characterization of the Sorbent	Adsorbed Amount/ Interactions	Reference
MIL-101(Cr)(F)	Activated at 150°C in air	Benzene	XRD, TEM, SEM, N_2 sorption, TGA, FTIR	Capacity exceeds that of activated carbon	Jhung et al. (2007)
MIL-101(Cr)	HF-free, stabilized with NH_4F; activated at 100°C in vacuum	Thiophene	XRD, N_2 sorption	$Q_0 = 6.3$ mg/g; polar interactions	Bhadra et al. (2015)
MIL-101(Cr)	Above	Pyrrole	Above	$Q_0 = 12$ mg/g; polar interactions	Bhadra et al. (2015)
MIL-101(Cr)	Above	Nitrobenzene	Above	$Q_0 = 33$ mg/g; polar interactions	Bhadra et al. (2015)
MIL-101(Cr)(F)	Activated at 150°C in air	Bisphenol A	XRD, N_2 sorption, BET, BJH	Langmuir constant $Q_0 = 252.5$ mg/g	Qin et al. (2015)

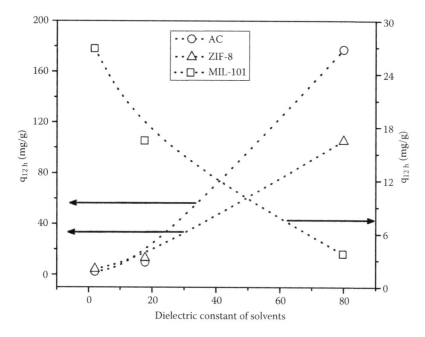

FIGURE 7.1   Effect of dielectric constant of solvents on the adsorbed quantity of Th (q_{12h}) over AC, MIL-101, and ZIF-8. The dotted lines were added to show the general tendency of the results. (Reprinted with permission from Bhadra, B.N. et al., *J. Phys. Chem. C*, 119(47), 26620–26627. Copyright 2015 American Chemical Society.)

TABLE 7.2
Adsorption of Perfluorooctanoic Acid (PFOA) on Pure versus Chemically Modified MIL-101(Cr) in Water

MOF	Details of MOF	Characterization of the Sorbent	Adsorbed Amount/ Adsorption Interactions	Reference
MIL-101(Cr)	F-free; dried in vacuum at 100°C	XRD, SEM, FTIR, XPS, N$_2$ sorption, ζ-potentials	Langmuir constant $q_m = 1.11$ mmol/g; ionic	Liu et al. (2015)
MIL-101(Cr)-DMEN	From MIL-101(Cr)	Above	$q_m = 1.29$ mmol/g; ionic	Liu et al. (2015)
MIL-101(Cr)-QDMEN	Methylation of MIL-101(Cr)-DMEN	Above	$q_m = 1.82$ mmol/g; ionic	Liu et al. (2015)
MIL-101 (Cr)-NH$_2$	From H$_2$N-BDC	Above	$q_m = 0.70$ mmol/g; ionic	Liu et al. (2015)
MIL-101(Cr)-NMe$_3$	Methylation of MIL-101(Cr)-NH$_2$	Above	$q_m = 1.19$ mmol/g; ionic	Liu et al. (2015)

derivative MIL-101-QDMEN was prepared through the reaction of MIL-101-DMEN with methyl triflate in dichloromethane (Liu et al. 2015). In the alternative method of synthesis from modified precursors, MIL-101(Cr)-NH_2 was synthesized from 2-amino-BDC precursor and $Cr(NO_3)_3$ and dried, and then quaternized MIL-101-NMe_3 was prepared from MIL-101(Cr)-NH_2 via the reaction with methyl triflate. Finally, both quaternized MIL-101-QDMEN and MIL-101-NMe_3 were acidified with HCl to obtain the corresponding anion-exchanged modified MOFs in their Cl⁻ form.

PFOA is a perfluorinated carboxylic acid frequently used as a chemically stable surfactant, including its industrial-scale applications in the polymerization of fluoropolymers in emulsion. Since the 1940s, PFOA has been manufactured in industrial quantities. In the past, PFOA was widely used in the production of polytetrafluoroethylene (Teflon). Like many fluoroorganic compounds, PFOA is very stable in wastewater and in soil. The toxicology studies on animals indicate that PFOA can cause several types of tumors and neonatal death and may have toxic effects on the immune, liver, and endocrine systems, and there is evidence of its toxicity to humans as well (Steenland et al. 2010). Due to the electron-withdrawing effects of fluorine atoms in the CF_2 chain, PFOA is a strong organic acid (Vierke et al. 2013). Therefore, for adsorption of PFOA in water by mesoporous MOFs, the ionic mechanism has been contemplated, which would require positively charged functional groups in the MOF sorbent.

The adsorption of PFOA in water resulted in the adsorption capacity of 1.19 and 1.82 mmol/g for modified MIL-101(Cr)-NMe_3 and MIL-101(Cr)-QDMEN (Liu et al. 2015). This significantly exceeded the adsorption capacity of the respective amino-containing MOFs in the same study. When the pH was increased within the range pH = 2–9, the adsorbed amount of PFOA steadily decreased. Since PFOA is a relatively strong organic acid (Goss 2008) with $pK_a \sim -0.5$, the change in the adsorbed amount because of pH change was concluded to be not due to the dissociation of POA. The ζ-potentials of quaternized and aminated MOFs were found to be higher than those of MIL-101(Cr), which explains an enhanced adsorption of PFOA in its anionic form. However, the ζ-potentials significantly decreased in the presence of Na+ as a counter-ion in the quaternized MOFs, which explains the more significant decrease in the adsorbed amounts of PFOA on quaternized MOFs versus nonquaternized ones. After three adsorption/desorption cycles, ca. 90% adsorption capacity toward PFOA was retained. The Fourier transform infrared (FTIR) spectroscopy study indicated the presence of spectral bands due to the NH and CH groups in MIL-101(Cr)-QDMEN after the fourth adsorption/desorption cycle, which suggests that the quaternary amine group was stable during repeated regeneration of the sorbent (Liu et al. 2015).

MIL-101(Cr)(F) is well-known to be stable in liquid water at room temperature, so this MOF is promising for adsorption-based applications. Other mesoporous MOFs were also studied for the adsorption of miscellaneous organic compounds in water (Table 7.3).

MIL-100(Fe), MIL-100(Cr), and NH_2-MIL-101(Al) were tested for the adsorption of phenol and p-nitrophenol (PNP) from water (Liu et al. 2014). The adsorption capacities for phenol were found to be similar on the three MOFs studied. It was then concluded that the metal cations in the MOFs were not the major adsorption

TABLE 7.3

Adsorption of Miscellaneous Organic Compounds in Water on Mesoporous MOFs Other Than MIL-101(Cr)

MOF	Details of the MOF	Adsorbate	Adsorbed Amount/ Adsorption Interactions	Reference
MIL-100(Fe)	Activated at 423 K in vacuum	Phenol; p-nitrophenol (PNP)	H-bonding with PNP	Liu et al. (2014)
MIL-100(Cr)	Above	Phenol (PH); PNP	H-bonding with PNP	Liu et al. (2014)
NH$_2$-MIL-101(Al)	Above	Phenol (PH); PNP	Langmuir constants PH: q$_e$ = 8.065 mmol/g; PNP: q$_e$ = 1.385 mmol/g; H-bonding with PNP	Liu et al. (2014)
MIL-100(Fe)(F)	Activated at 80°C in vacuum	Bisphenol A	Langmuir constant Q$_0$ = 55.6 mg/g	Qin et al. (2015)

sites. Interestingly, NH$_2$-MIL-101(Al) demonstrated much higher capacity for PNP versus the other MOFs. In this mechanistic study, salicylic acid with pK_a close to that of PNP had about the same adsorbed amounts on the three MOFs. Therefore, the acid–base interaction of the PNP with the MOFs was thought not to be the dominant mechanism. The higher adsorption capacity of PNP versus phenol has been explained by the strong hydrogen bonding involving the amino groups in NH$_2$-MIL-101(Al) (Liu et al. 2014, Figure 7.2).

Bisphenol A (BPA) (Figure 7.3), or [2,2-bis(4-hydroxyphenyl)propane], is widely used in manufacturing epoxy and polycarbonate plastics, including baby bottles, lining of food cans, and dental sealants (Staples et al. 1998).

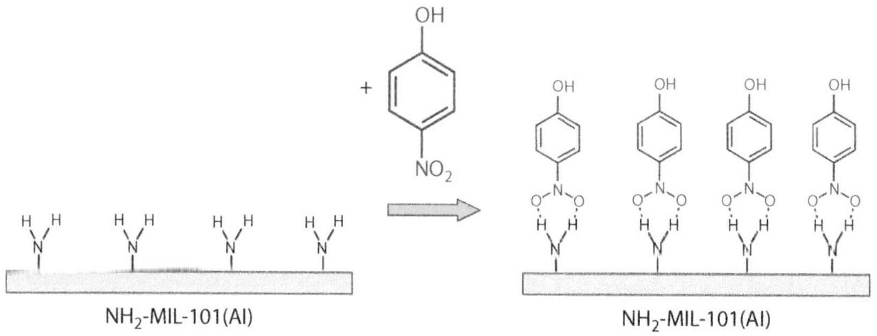

FIGURE 7.2 The possible mechanism of PNP adsorption on NH$_2$-MIL-101(Al) by hydrogen bonding. (Reprinted with permission from Liu, B. et al., *J. Chem. Eng. Data*, 59(5), 1476–1482. Copyright 2014 American Chemical Society.)

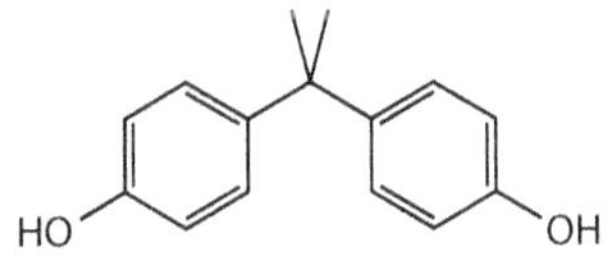

FIGURE 7.3   Bisphenol A (BPA).

BPA was shown to be one of the most abundant endocrine-disrupting chemicals (EDCs) that may cause negative effects on the endocrine systems of humans and wildlife. The adsorptive removal of BPA from water has been studied on MIL-101(Cr) and MIL-100(Fe) (Qin et al. 2015). The adsorbed amount of 156.4 mg/g at C_0 = 120 ppm on MIL-101(Cr) was much higher than that of 26.0 mg/g on MIL-100(Fe) under the same conditions. The size of the BPA molecule is ca. 9.4 × 5.9 × 4.8 Å, which is smaller than the large aperture in MIL-101(Cr) at 12.6 Å, but is comparable to the aperture of ca. 8.6 Å in MIL-100(Fe). These sterical considerations were used to explain the much higher adsorbed amounts of BPA on MIL-101(Cr). In addition, the adsorbed amount of BPA on MIL-101(Cr), MIL-100(Fe), and AC was compared, and the adsorbed amount was higher when the total surface area was higher. At neutral pH, hydrogen bonding between BPA and MIL-101(Cr) and π–π interactions were thought to be responsible for the adsorption. At pH > 10, the adsorption capacity of MIL-100(Fe) was very low, possibly due to the dissolution of this MOF (Qin et al. 2015).

Many polyaromatic hydrocarbons (PAHs) are highly toxic (Djomo et al. 2004) and can accumulate in water and soil. Therefore, methods for their removal and chemical analysis in water and soil need to be developed. The nonselective adsorption of several PAHs, naphthalene (Nap), acenaphthene (Ace), anthracene (Ant), fluoranthene (FluA), pyrene (Pyr), and benz[a]anthracene (BaA) has been studied on the magnetic composite material (Huo and Yan 2012) formed by iron oxide with MIL-101(Cr). To prepare the composite, in situ magnetization of MIL-101(Cr) was undertaken by mixing with silica-coated microparticles of Fe_3O_4 in solution under sonication. In the same suspension, magnetic solid-phase extraction of polycyclic aromatic hydrocarbons from water on magnetic MIL-101(Cr) composite was studied (Huo and Yan 2012). The solid-phase extraction was utilized for the separation and quantitative determination of PAHs by the high-performance liquid chromatography (HPLC) method. The detection limit was 2.8–27.2 ng/L, and the quantification limit was 6.3–87.7 ng/L. It was assumed that the combination of the nonspecific hydrophobic interactions, π–π interactions between the PAHs and BDC linker of the MOF, and π-complexation between the PAHs and the Lewis acid sites in the pores of MIL-101(Cr) may play a role in the adsorption; however, the dominating mechanism is unknown.

Chlorinated aromatic compounds are widely used as herbicides, and they are the major toxic contaminants of water and soil worldwide (Pereira et al. 1996, Muller et al. 2008). To remove toxic and chemically stable chlorinated aromatic compounds from water and soil, various methods have been explored. The available protocols include those based on chemically aggressive, expensive, or energy-demanding methods, for example, sonochemistry combined with photocatalysis (Peller et al. 2003), electrocatalysis using Pt group metals (Tsyganok and Otsuka 1999), extraction by supercritical carbon dioxide (Laitinen et al. 1994), and chemical reaction with sodium metal (Pittman and He 2002). The adsorptive removal of chlorinated aromatic compounds from water by water-stable mesoporous MOFs seems to be an effective approach, since mesoporosity favors the adsorption of larger molecules (Kim et al. 2013), which feature the highest toxicity. Yet, only one paper reports the adsorptive removal of chlorinated aromatic compounds by mesoporous MOFs; however, the solvent was an aromatic compound (Kim et al. 2013). It is rather surprising that the adsorption of toxic chlorinated aromatic compounds from water has not been reported on mesoporous MOFs. This research avenue is definitely worthy of exploration.

REFERENCES

Bhadra, B. N., K. H. Cho, N. A. Khan, D. Y. Hong, and S. H. Jhung. 2015. Liquid-phase adsorption of aromatics over a metal-organic framework and activated carbon: Effects of hydrophobicity/hydrophilicity of adsorbents and solvent polarity. *Journal of Physical Chemistry C* 119(47):26620–26627.

Djomo, J. E., A. Dauta, V. Ferrier, J. F. Narbonne, A. Monkiedje, T. Njine, and P. Garrigues. 2004. Toxic effects of some major polyaromatic hydrocarbons found in crude oil and aquatic sediments on *Scenedesmus subspicatus*. *Water Research* 38(7):1817–1821.

Goss, K.-U. 2008. The pK_a values of PFOA and other highly fluorinated carboxylic acids. *Environmental Science & Technology* 42(2):456–458.

Hong, D. Y., Y. K. Hwang, C. Serre, G. Ferey, and J. S. Chang. 2009. Porous chromium terephthalate MIL-101 with coordinatively unsaturated sites: Surface functionalization, encapsulation, sorption and catalysis. *Advanced Functional Materials* 19(10):1537–1552.

Huo, S. H. and X. P. Yan. 2012. Facile magnetization of metal-organic framework MIL-101 for magnetic solid-phase extraction of polycyclic aromatic hydrocarbons in environmental water samples. *Analyst* 137(15):3445–3451.

Jhung, S. H., J. H. Lee, J. W. Yoon, C. Serre, G. Ferey, and J. S. Chang. 2007. Microwave synthesis of chromium terephthalate MIL-101 and its benzene sorption ability. *Advanced Materials* 19(1):121–124.

Kim, H. Y., S. N. Kim, J. Kim, and W. S. Ahn. 2013. Liquid phase adsorption of selected chloroaromatic compounds over metal organic frameworks. *Materials Research Bulletin* 48(11):4499–4505.

Laitinen, A., A. Michaux, and O. Aaltonen. 1994. Soil cleaning by carbon-dioxide extraction—A review. *Environmental Technology* 15(8):715–727.

Liu, B., F. Yang, Y. Zou, and Y. Peng. 2014. Adsorption of phenol and p-nitrophenol from aqueous solutions on metal–organic frameworks: Effect of hydrogen bonding. *Journal of Chemical & Engineering Data* 59(5):1476–1482.

Liu, K., S. Y. Zhang, X. Y. Hu, K. Y. Zhang, A. Roy, and G. Yu. 2015. Understanding the adsorption of PFOA on MIL-101(Cr)-based anionic-exchange metal-organic frameworks: Comparing DFT calculations with aqueous sorption experiments. *Environmental Science & Technology* 49(14):8657–8665.

Muller, B., M. Berg, Z. P. Yao, X. F. Zhang, D. Wang, and A. Pfluger. 2008. How polluted is the Yangtze River? Water quality downstream from the Three Gorges Dam. *Science of the Total Environment* 402(2–3):232–247.

Peller, J., O. Wiest, and P. V. Kamat. 2003. Synergy of combining sonolysis and photocatalysis in the degradation and mineralization of chlorinated aromatic compounds. *Environmental Science & Technology* 37(9):1926–1932.

Pereira, W. E., J. L. Domagalski, F. D. Hostettler, L. R. Brown, and J. B. Rapp. 1996. Occurrence and accumulation of pesticides and organic contaminants in river sediment, water and clam tissues from the San Joaquin River and tributaries, California. *Environmental Toxicology and Chemistry* 15(2):172–180.

Pittman, C. U. and J. B. He. 2002. Dechlorination of PCBs, CAHs, herbicides and pesticides neat and in soils at 25 degrees C using Na/NH$_3$. *Journal of Hazardous Materials* 92(1):51–62.

Qin, F. X., S. Y. Jia, Y. Liu, H. Y. Li, and S. H. Wu. 2015. Adsorptive removal of bisphenol A from aqueous solution using metal-organic frameworks. *Desalination and Water Treatment* 54(1):93–102.

Staples, C. A., P. B. Dome, G. M. Klecka, S. T. Oblock, and L. R. Harris. 1998. A review of the environmental fate, effects, and exposures of bisphenol A. *Chemosphere* 36(10):2149–2173.

Steenland, K., T. Fletcher, and D. A. Savitz. 2010. Epidemiologic evidence on the health effects of perfluorooctanoic acid (PFOA). *Environmental Health Perspectives* 118(8):1100–1108.

Tsyganok, A. I. and K. Otsuka. 1999. Selective dechlorination of chlorinated phenoxy herbicides in aqueous medium by electrocatalytic reduction over palladium-loaded carbon felt. *Applied Catalysis B: Environmental* 22(1):15–26.

Vierke, L., U. Berger, and I. T. Cousins. 2013. Estimation of the acid dissociation constant of perfluoroalkyl carboxylic acids through an experimental investigation of their water-to-air transport. *Environmental Science & Technology* 47(19):11032–11039.

8 Adsorption of Inorganic Ions on Mesoporous MOFs from Water

Research on the adsorption of inorganic ions on mesoporous metal–organic frameworks (MOFs) in the aqueous phase has been published rarely. The reason, apparently, is the higher cost involved in the adsorption of toxic organic compounds such as synthetic dyes, pesticides, and pharmaceuticals. Table 8.1 shows published data on adsorption of cations of metals on MIL-101 MOFs.

The MIL-101 sorbent containing the thiol group in the linker was prepared by the multistep postmodification procedure (Liu et al. 2014). First, MIL-101(Cr) was synthesized from $Cr(NO_3)_3$, terephthalic acid, and hydrofluoric acid by the well-established procedure (Férey et al. 2005). Then, the obtained MIL-101(Cr) was nitrated to introduce the $-NO_2$ group to the phenyl ring of the linker. Then, the obtained MIL-101(Cr)–NO_2 was reduced with $SnCl_2$ to obtain H_2N–MIL-101(Cr) denoted here as Cr-MIL-101–NH_2. Next, postsynthetic modification (PSM) of the resultant Cr-MIL-101–NH_2 with $NaNO_2$ conducted as was reported earlier (Jiang et al. 2012) resulted in the intermediate modified MOF denoted Cr-MIL-101–N_2BF_4 (Liu et al. 2014). The further reaction of Cr-MIL-101–N_2BF_4 with substituted alkenes resulted in the substituted MOFs denoted Cr-MIL-101–A1 and Cr-MIL-101–A5 (Figure 8.1).

In MIL-101–A1, the R radical is the phenyl group, while in MIL-101–A5 the R radical contains both the olefin and the ester functional groups (Liu et al. 2014). Cr-MIL-101–A1 and Cr-MIL-101–A5 were tested as sorbents in the removal of Hg^{2+} from water containing 2% HNO_3 where Hg^{2+} cations were present at the initial concentration $C_0 = 10$ ppm. After 6 h of adsorption, Cr-MIL-101–A1 and Cr-MIL-101–A5 sorbents caused a decrease in the concentration of Hg^{2+} down to 1.03 and 0.96 ppm, respectively. This has been considered by the authors as rather moderate adsorption capacity for Hg^{2+}, due to the weak interactions between Hg^{2+} as the Lewis acid and the π-electron systems of the linkers in the MOF as the Lewis base (Liu et al. 2014). Stronger affinity of the functional group in the postsynthetically modified MOF toward Hg^{2+} was needed. The so-called "tandem PSM method" resulted in the new MOF denoted Cr-MIL-101–AS, which carries the desired terminal thiol group (Figure 8.2).

The obtained Cr-MIL-101–AS demonstrated very good adsorption capacity for Hg^{2+} in water, with the removal efficiency for mercury at 99.3% at $C_0 = 10$ ppm and 93.4% at $C_0 = 0.1$ ppm. Such a high adsorption capacity for Hg^{2+} was attributed to the interactions between the mercury cation and the thiol group in Cr-MIL-101–AS. The "spent" Cr-MIL-101–AS–Hg sorbent was reused twice without a significant loss of activity, with desorption of Hg^{2+} using the solution of thiourea in 1 M hydrochloric acid (Liu et al. 2014).

TABLE 8.1

Adsorption of Metal Cations on MIL-101 from Water

MOF	Details of the MOF	Adsorbate	Adsorbed Amount/ Interactions	Reference
Cr-MIL-101–A1	Liu et al. (2014)	Hg^{2+}	89.8% for $C_0 = 10$ ppm/coordination	Liu et al. (2014)
Cr-MIL-101–A5	Liu et al. (2014)	Hg^{2+}	90.3% for $C_0 = 10$ ppm/coordination	Liu et al. (2014)
Cr-MIL-101–AS	Figure 8.2	Hg^{2+}	93.4% for $C_0 = 0.1$ ppm/coordination	Liu et al. (2014)
MIL-101(Cr)		$UO_2(NO_3)_2$		Bai et al. (2015)
MIL-101–NH$_2$	MIL-101(Cr)-NH$_2$	$UO_2(NO_3)_2$		Bai et al. (2015)
MIL-101–ED	Activated, then PSM with ethylenediamine	$UO_2(NO_3)_2$		Bai et al. (2015)
MIL-101–DETA	Activated, then PSM with diethylenetriamine	$UO_2(NO_3)_2$		Bai et al. (2015)
MIL-101(Cr)(F)	Activated at 75°C overnight	UO_2^{2+}	34.38 mg/g at $C_0 =$ 100 ppm	Zhang et al. (2015)
ED-MIL-101(Cr)(F)	Activated at RT overnight	UO_2^{2+}	200 mg/g at $C_0 =$ 100 ppm	Zhang et al. (2015)

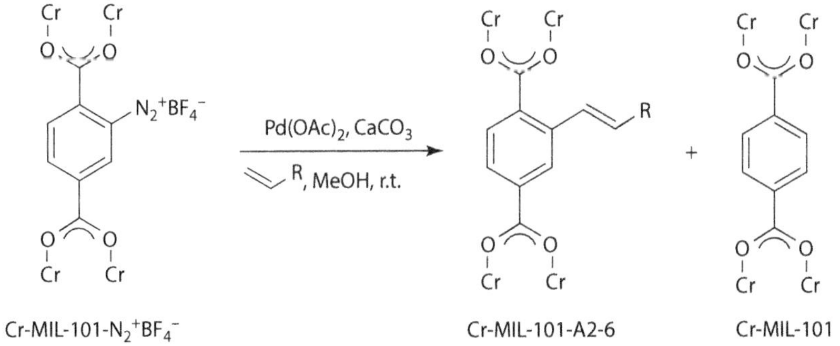

FIGURE 8.1 The palladium-catalyzed carbon–carbon functionalization of Cr-MIL-101. (Modified from Liu, T., Che, J.X., Hu, Y.Z., Dong, X.W., Liu, X.Y., and Che, C.M.: Alkenyl/thiol-derived metal-organic frameworks (MOFs) by means of postsynthetic modification for effective mercury adsorption. *Chem. Eur. J.* 2014. 20(43). 14090–14095. Copyright Wiley-VCH Verlag GmbH & Co. KGaA. Reproduced with permission.)

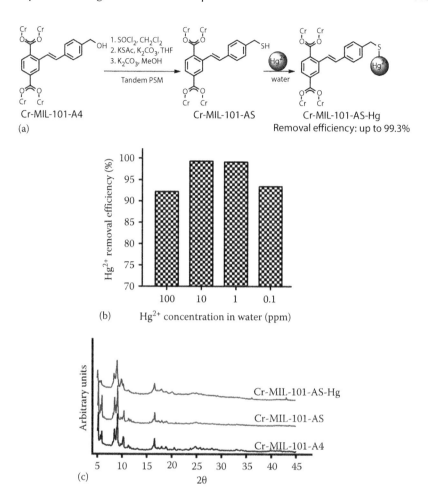

FIGURE 8.2 (a) The tandem post-synthetic modification (PSM) of Cr-MIL-101–A4 to Cr-MIL-101–AS and its application in mercury adsorption. (b) The high mercury adsorption efficiency of Cr-MIL-101–AS in water. (c) Powder x-ray diffraction (PXRD) measurements of Cr-MIL-101–A4, Cr-MIL-101–AS, and Hg-adsorbed Cr-MIL-101–AS–Hg. (Reproduced from Liu, T., Che, J.X., Hu, Y.Z., Dong, X.W., Liu, X.Y., and Che, C.M.: Alkenyl/thiol-derived metal-organic frameworks (MOFs) by means of postsynthetic modification for effective mercury adsorption. *Chem. Eur. J.* 2014. 20(43). 14090–14095. Copyright Wiley-VCH Verlag GmbH & Co. KGaA. Reproduced with permission.)

With the increasing demand for energy and the need to limit the use of fossil fuels, the utilization of nuclear power is predicted to increase. Uranium is the major fuel in nuclear reactors. The world's oceans contain approximately 4.5 billion tons of uranium, which is a thousand times more than the estimated amount of uranium present in ores worldwide. Therefore, the extraction of uranium from seawater is being actively explored. In addition, with the further development of nuclear power infrastructure, large amounts of radioactive wastewaters containing uranium will be produced, so it is desirable to have suitable methods for water purification from uranium. Adsorption of uranium from water is a promising method (Li and Zhang 2012) among the other methods that include ion exchange and extraction.

Chemically modified sorbents MIL-101–NH$_2$, MIL-101–ED (ED = ethylenediamine), and MIL-101–DETA (DETA = diethylenetriamine) were synthesized from MIL-101(Cr)(F) by PSM, characterized, and tested in the adsorption of uranyl nitrate UO$_2$(NO$_3$)$_2$ from water at room temperature (Bai et al. 2015). The highest adsorption capacity was achieved with MIL-101–DETA at C$_0$ = 350 mg/L and at pH ~ 5.5. The adsorbed amounts followed the given order: MIL-101–DETA > MIL-101–ED > MIL-101–NH$_2$ > MIL-101(Cr)(F). The adsorption mechanism was explained as the coordination of U(VI) with the amino groups in the modified MOFs. Regeneration of the "spent" sorbents via desorption of U(VI) was achieved by lowering the pH ≤ 3, and >99% uranium was desorbed. This finding indicates protonation of the amino groups and destruction of their coordination complexes with adsorbed U(VI). The regenerated MIL-101–NH$_2$ did not lose its adsorption capacity, while about 30% loss of capacity was observed for the chemically more complex sorbents MIL-101–ED and MIL-101–DETA (Bai et al. 2015).

Similar work on the adsorption of uranyl ions UO$_2^{2+}$ on ED-functionalized MIL-101(Cr)(F) was reported (Zhang et al. 2015). By PSM, MIL-101(Cr)(F) was grafted through the coordination of ED on the coordinatively unsaturated site (CUS) of Cr(III), yielding several ED–MIL-101(Cr)(F) MOFs. The obtained ED–MIL-101(Cr)(F) MOFs were tested for the adsorption of U(VI) in the form of uranyl cations UO$_2^{2+}$ from water. ED–MIL-101(Cr)(F) demonstrated higher adsorption capacity than MIL-101(Cr)(F), although the Brunauer–Emmett–Teller (BET) total surface area was lower for the former. The pH was varied in the range pH = 2–9, and the maximum adsorption capacity was observed at pH = 4.5. This finding has been explained by the UO$_2^{2+}$ cations being the main species of uranium present in the range pH = 2–5; thus, UO$_2^{2+}$ are believed to directly interact with the amino groups in ED–MIL-101(Cr)(F). The adsorbed U(VI) was desorbed by setting the pH ≤ 2.0, apparently due to the protonation of amino groups and the resultant destruction of the coordination bonds formed by the UO$_2^{2+}$ ions. Indeed, according to X-ray absorption spectroscopy (XAS) measurements, the adsorbed U(VI) cations are coordinated to the amine groups in ED–MIL-101(Cr)(F), while physical adsorption occurred in MIL-101(Cr)(F). After four adsorption/regeneration cycles, about 58% decrease in the adsorption capacity was observed, possibly due to some loss of the ED functionality in ED–MIL-101(Cr)(F) via substitution with the water molecules under regeneration in acidic conditions. However, the structure of this MOF has been found to remain stable after sorbent regeneration (Zhang et al. 2015).

Among the inorganic anions present in the environment, arsenates and arsenites belong to the most toxic compounds. Arsenic in the As(V) and As(III) forms is considered to be one of the most toxic inorganic pollutants present in water. Arsenic is one of the major toxic by-products in metallurgy of nonferrous metals (Hopkin 1989) and is one of the few chemical elements whose production by far exceeds its consumption. In the past, arsenic found widespread use in inorganic pigments. Scheele's Green pigment is copper arsenite (CuHAsO$_3$), which was used in paints in the nineteenth century (Castro et al. 2004), but it was later prohibited due to its high toxicity. Copper(II) acetoarsenite also known as Paris Green is a greenish highly toxic powder that was widely used in the first half of the twentieth century as a pigment, as a colorant for fireworks, as rodenticide, and as insecticide. Arsenic is believed to have a carcinogenic effect (Martinez et al. 2011). Contemporary methods for the removal of arsenic from water are based on adsorption (Mohan and Pittman 2007). Water-stable MOFs are promising sorbents for removal of arsenic from water.

Fe–BTC MOF was prepared from the $FeCl_3$ precursor and 1,3,5-benzenetricar-boxylic acid in dimethylformamide (DMF) (Zhu et al. 2012); see Table 8.2.

By its chemical composition, this MOF is similar to fluoride-free MIL-100(Fe). Fe–BTC was tested in the adsorptive removal of As(V) from water using sodium arsenate Na_3AsSO_4 (Zhu et al. 2012). Fe–BTC demonstrated adsorption capacity >6-fold that of iron oxide nanoparticles and 36-fold that of commercial powder of iron oxide (Zhu et al. 2012), which is widely used to remove arsenic from water (Gallegos-Garcia et al. 2012). In water at pH < 6.9, the main form of arsenate is $H_2AsO_4^-$, while $HAsO_4^{2-}$ is the dominant form at pH > 7. In addition, at the very low pH < 2.3 and at very high pH, H_3AsO_4 and AsO_4^{3-}, respectively, are present. The adsorption efficiency of Fe–BTC for As(V) was found to be 96% in the range pH = 2–10 (Zhu et al. 2012). The highest adsorption efficiency of 98.2% was determined at pH = 4. However, in basic solutions at pH > 12, the removal efficiency of As(V) drastically decreased to 35.8% due to gradual dissolution of Fe–BTC. The pH of natural waters is in the range 6.0–8.5; therefore, Fe–BTC MOF seems to be suitable for large-scale removal of As(V) from water. The mechanism of adsorption has been investigated by Fourier transform infrared (FTIR) spectroscopy. The infrared (IR) peak at 824 cm^{-1} appeared after the adsorption of As(V) and it was assigned to the Fe–O–As group (Zhu et al. 2012).

An interesting study was published recently on the adsorption of iodide anion from water. MIL-101(Cr)–SO_3H was transformed to MIL-101(Cr)–SO_3Ag via ion exchange, namely, by interaction with water solution of $AgNO_3$, and it was then dried and characterized (Zhao et al. 2015). The adsorption capacity with MIL-101(Cr)–SO_3Ag toward iodide anion was 244.2 mg/g versus 94.1 mg/g obtained with MIL-101(Cr)–SO_3H.

To understand the mechanism of adsorption of iodide anion, measurements of the ζ-potential were conducted (Zhao et al. 2015). The surface of MIL-101(Cr)–SO_3Ag was found to remain negatively charged both before and after the adsorption of iodide anion. This suggests that electrostatic interactions did not play a major role. Instead, the increase in the adsorption capacity was attributed to the ion exchange with iodide adsorbate: MIL-101(Cr)–SO_3Ag + NaI → MIL-101(Cr)–SO_3Na + AgI (precipitate). This implies that MIL-101(Cr)–SO_3Ag is a "disposable" sorbent. Indeed, the sorbent regeneration procedure and the tests for regeneration were not provided (Zhao et al. 2015).

For practical applications, the selectivity of adsorption of ionic compounds in the presence of other ionic compounds is important. The studies of the selectivity of adsorption of metal cations in water on the mesoporous MOFs are summarized in Table 8.3.

The selectivity of adsorption of Pb^{2+} ions in water on MIL-101(Cr) and ED-grafted ED–MIL-101(Cr) was determined in the presence of Cu^{2+}, Zn^{2+}, Co^{2+}, and Ni^{2+} at the initial concentration $C_0 = 30$ mg/L (Luo et al. 2015). ED–MIL-101(Cr) demonstrated higher adsorption selectivity Pb(II) ions than MIL-101(Cr) (Figure 8.3).

TABLE 8.2
Adsorption of Inorganic Anions on Mesoporous MOFs in Water

MOF	Details of the MOF	Adsorbate	Characterization of the Sorbent	Adsorbed Amount/ Interactions	Reference
Fe–BTC	HF-free synthesis; activated at 60°C in vacuum	Na_3AsO_4	XPS, SEM, TEM, FTIR, TGA	Maximum capacity q_m = 12.287 mg/g	Zhu et al. (2012)
MIL-101(Cr)-SO$_3$H	Solvothermal reaction of CrO$_3$, monosodium 2-sulfoterephthalic acid and HCl in water	NaI	XRD, BET, TGA, FESEM, FTIR, XPS, EDX, ζ-potential	94.1 mg/g iodide	Zhao et al. (2015)
MIL-101(Cr)-SO$_3$Ag	Synthesis by ion exchange with AgNO$_3$, dried at 358 K in vacuum	NaI	Above	244.2 mg iodide/g sorbent/ion exchange and formation of AgI	Zhao et al. (2015)

Abbreviation: BTC, benzenetricarboxylic acid.

TABLE 8.3

Selectivity of Adsorption of Metal Cations on Mesoporous MOFs in Water

MOF	Details of the MOF	Adsorbate	Competing Ions	Adsorption Selectivity of Target Analyte	Reference
MIL-101(Cr)(F)	Characterized by XRD, FTIR, SEM, TEM, BET, XPS	Pb^{2+}	Cu^{2+}, Zn^{2+}, Co^{2+}, Ni^{2+}	Distribution coefficient $K_d = 0.319$ for Pb^{2+}	Luo et al. (2015)
ED-MIL-101(Cr)(F)	Above	Pb^{2+}	Cu^{2+}, Zn^{2+}, Co^{2+}, Ni^{2+}	$K_d = 1.177$ for Pb^{2+}	Luo et al. (2015)
MIL-101–DETA	Activated and postmodified with DETA, diethylenetriamine	$UO_2(NO_3)_2$	Co^{2+}, Ni^{2+}, Zn^{2+}, Sr^{2+}, La^{3+}, Nd^{3+}, Sm^{3+}, Gd^{3+}, Yb^{3+}	Highest selectivity at pH = 5; distribution coefficient $K_d > 6000$ mL/g at $C_0 = 0.5$ mmol/L	Bai et al. (2015)

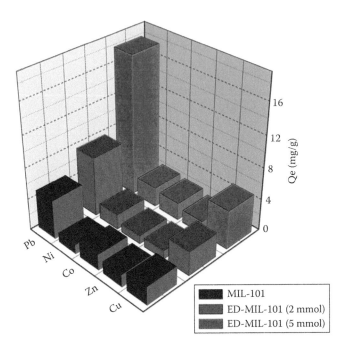

FIGURE 8.3 Selectivity of MIL-101, ED–MIL-101 (2 mmol), and ED–MIL-101 (5 mmol). (Reprinted with permission from Luo, X.B., Ding, L., and Luo, J.M., Adsorptive removal of Pb(II) ions from aqueous samples with amino-functionalization of metal-organic frameworks MIL-101(Cr), *J. Chem. Eng. Data*, 60(6), 1732–1743. Copyright 2015 American Chemical Society.)

The high adsorption selectivity was explained by the preferential coordination of Pb^{2+} to the amino groups (Figure 8.4) compared to the interfering cations (Luo et al. 2015).

The adsorption of U(VI) present as $UO_2(NO_3)_2$ on MIL-101–DETA has been tested in the presence of interfering cations Co, Ni, Zn, Sr, La, Nd, Sm, Gd, and Yb at the initial concentration $C_0 = 0.5$ mmol/L for each cation (Bai et al. 2015). At pH = 5.5, the adsorption capacity for U(VI) did not significantly change in the presence of interfering cations, which means that adsorption selectivity was achieved. This is illustrated by the relatively high value of the distribution coefficient for U(VI), with $K_d > 6000$ mL/g at pH = 5.5, which is close to the typical pH range of natural waters.

When high adsorption selectivity is not achieved or not desired, nonselective adsorption of several similar ions and their "preconcentration" on the MOF sorbent can be used. Mesoporous MOF can be used both as a sorbent and as the mesoporous structural template for another sorbent (inorganic, organic, or the hybrid material). The novel magnetic metal–MOF composite material [(Fe$_3$O$_4$–En)/MIL-101(Fe)] has been synthesized (Babazadeh et al. 2015a; Figure 8.5). In Figure 8.5, AEAPTMS denotes N-(2-aminoethyl)-3-(aminopropyl)trimethoxysilane and EN denotes ethylenediamine.

The composite sorbent [(Fe$_3$O$_4$–En)/MIL-101(Fe)] has been tested to preconcentrate small amounts of Cd^{2+}, Pb^{2+}, Zn^{2+}, and Cr^{3+} ions from water (Babazadeh et al. 2015a).

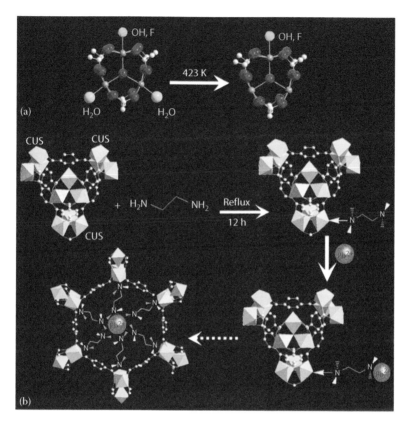

FIGURE 8.4 (a) Generation of coordinatively unsaturated sites from chromium trimers in MIL-101 after vacuum treatment at 423 K for 12 h; (b) adsorption principle of amino-functionalized MIL-101 for Pb(II) ions. (Reprinted with permission from Luo, X.B., Ding, L., and Luo, J.M., Adsorptive removal of Pb(II) ions from aqueous samples with amino-functionalization of metal-organic frameworks MIL-101(Cr), *J. Chem. Eng. Data*, 60(6), 1732–1743. Copyright 2015 American Chemical Society.)

The adsorption capacity was found to be 155 mg/g for cadmium, 198 mg/g for lead, 164 mg/g for zinc, and 173 mg/g for chromium. The limits of detection (LODs) were found to be 0.15, 0.8, 0.2, and 0.5 ng/mL for Cd^{2+}, Pb^{2+}, Zn^{2+}, and Cr^{3+}, respectively. This magnetic nanocomposite material was applied to extract trace amounts of these heavy metal ions from several vegetables, including leek, fenugreek, parsley, radish, radish leaves, beetroot leaves, garden cress, basil, and coriander.

Another study that utilized similarly prepared nanocomposite [(Fe$_3$O$_4$ dipyridylamine)/MIL-101(Fe)] to preconcentrate the trace amounts of Cd^{2+}, Pb^{2+}, Co^{2+}, and Ni^{2+} cations from fish samples was reported by the same group (Babazadeh et al. 2015b).

Research on preconcentration of metal cations from water on mesoporous MOFs and composites of mesoporous MOFs with other suitable sorbents is ongoing. As in a similar research area of using MOFs as stationary phases in chromatography

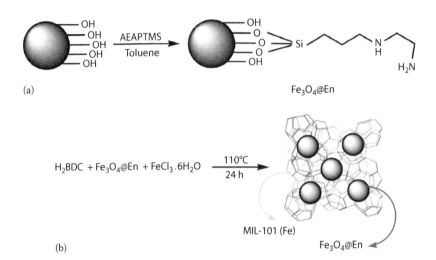

(a)   Fe$_3$O$_4$@En

H$_2$BDC + Fe$_3$O$_4$@En + FeCl$_3$.6H$_2$O $\xrightarrow[24\ h]{110°C}$

MIL-101 (Fe)

(b)   Fe$_3$O$_4$@En

FIGURE 8.5 (a) A schematic diagram of Fe$_3$O$_4$ functionalization by En. (b) The schematic illustration of synthesized magnetic MOF nanocomposite. (From Babazadeh, M., Hosseinzadeh-Khanmiri, R., Abolhasani, J., Ghorbani-Kalhor, E., and Hassanpour, A., Solid phase extraction of heavy metal ions from agricultural samples with the aid of a novel functionalized magnetic metal-organic framework, *RSC Adv.*, 5(26), 19884–19892, 2015. Reproduced by permission of The Royal Society of Chemistry.)

(e.g., Yu et al. 2013), mechanistic studies with multiple adsorbates are difficult to perform. Normally, one would select a "prototype" adsorbate for which the experimental study on the adsorption mechanism can be conducted, and the other adsorbate species in the preconcentration study are considered to behave similarly in the adsorption process.

REFERENCES

Babazadeh, M., R. Hosseinzadeh-Khanmiri, J. Abolhasani, E. Ghorbani-Kalhor, and A. Hassanpour. 2015a. Solid phase extraction of heavy metal ions from agricultural samples with the aid of a novel functionalized magnetic metal-organic framework. *RSC Advances* 5(26):19884–19892.

Babazadeh, M., R. Hosseinzadeh-Khanmiri, J. Abolhasani, E. Ghorbani-Kalhor, and A. Hassanpour. 2015b. Synthesis and application of a novel functionalized magnetic metal-organic framework sorbent for determination of heavy metal ions in fish samples. *Bulletin of the Chemical Society of Japan* 88(6):871–879.

Bai, Z. Q., L. Y. Yuan, L. Zhu, Z. R. Liu, S. Q. Chu, L. R. Zheng, J. Zhang, Z. F. Chai, and W. Q. Shi. 2015. Introduction of amino groups into acid-resistant MOFs for enhanced U(VI) sorption. *Journal of Materials Chemistry A* 3(2):525–534.

Castro, K., M. Perez-Alonso, M. D. Rodriguez-Laso, and J. M. Madariaga. 2004. Pigment analysis of a wallpaper from the early 19th century: Les Monuments de Paris. *Journal of Raman Spectroscopy* 35(8–9):704–709.

Férey, G., C. Mellot-Draznieks, C. Serre, F. Millange, J. Dutour, S. Surblé, and I. Margiolaki. 2005. A chromium terephthalate-based solid with unusually large pore volumes and surface area. *Science* 309(5743):2040–2042.

Gallegos-Garcia, M., K. Ramírez-Muñiz, and S. Song. 2012. Arsenic removal from water by adsorption using iron oxide minerals as adsorbents: A review. *Mineral Processing and Extractive Metallurgy Review* 33(5):301–315.

Hopkin, W. 1989. The problem of arsenic disposal in non-ferrous metals production. *Environmental Geochemistry and Health* 11(3-4):101–112.

Jiang, D., L. L. Keenan, A. D. Burrows, and K. J. Edler. 2012. Synthesis and post-synthetic modification of MIL-101(Cr)-NH$_2$ via a tandem diazotisation process. *Chemical Communications* 48(99):12053–12055.

Li, J. and Y. Zhang. 2012. Remediation technology for the uranium contaminated environment: A review. *Procedia Environmental Sciences* 13:1609–1615.

Liu, T., J. X. Che, Y. Z. Hu, X. W. Dong, X. Y. Liu, and C. M. Che. 2014. Alkenyl/thiol-derived metal-organic frameworks (MOFs) by means of postsynthetic modification for effective mercury adsorption. *Chemistry—A European Journal* 20(43):14090–14095.

Luo, X. B., L. Ding, and J. M. Luo. 2015. Adsorptive removal of Pb(II) ions from aqueous samples with amino-functionalization of metal-organic frameworks MIL-101(Cr). *Journal of Chemical and Engineering Data* 60(6):1732–1743.

Martinez, V. D., E. A. Vucic, D. D. Becker-Santos, L. Gil, and W. L. Lam. 2011. Arsenic exposure and the induction of human cancers. *Journal of Toxicology* 2011. Article ID 431287:1–13.

Mohan, D. and C. U. Pittman Jr. 2007. Arsenic removal from water/wastewater using adsorbents—A critical review. *Journal of Hazardous Materials* 142(1–2):1–53.

Yu, Y. B., Y. Q. Ren, W. Shen, H. M. Deng, and Z. Q. Gao. 2013. Applications of metal-organic frameworks as stationary phases in chromatography. *TrAc Trends in Analytical Chemistry* 50:33–41.

Zhang, J. Y., N. Zhang, L. J. Zhang, Y. Z. Fang, W. Deng, M. Yu, Z. Q. Wang, L. Li, X. Y. Liu, and J. Y. Li. 2015. Adsorption of uranyl ions on amine-functionalization of MIL-101(Cr) nanoparticles by a facile coordination-based post-synthetic strategy and x-ray absorption spectroscopy studies. *Scientific Reports* 5. Article number 13514:1–10.

Zhao, X. D., X. Han, Z. J. Li, H. L. Huang, D. H. Liu, and C. L. Zhong. 2015. Enhanced removal of iodide from water induced by a metal-incorporated porous metal-organic framework. *Applied Surface Science* 351:760–764.

Zhu, B.-J., X.-Y. Yu, Y. Jia, F.-M. Peng, B. Sun, M.-Y. Zhang, T. Luo, J.-H. Liu, and X.-J. Huang. 2012. Iron and 1,3,5-benzenetricarboxylic metal–organic coordination polymers prepared by solvothermal method and their application in efficient As(V) removal from aqueous solutions. *Journal of Physical Chemistry C* 116(15):8601–8607.

9 Adsorption of Aromatic N-Heterocyclic Compounds from Liquid Fossil Fuels

9.1 AROMATIC N-HETEROCYCLIC COMPOUNDS IN FOSSIL FUELS

Heterocyclic compounds have found wide and diverse applications. One can distinguish between heterocyclic compounds with oxygen, nitrogen, and sulfur atoms in the ring; the other heterocyclic compounds are much less common. Heterocyclic compounds most abundantly present in nature are, probably, the aromatic nitrogen and aromatic sulfur compounds.

Aromatic nitrogen-heterocyclic compounds, aka N-heterocyclic compounds, are abundantly present in petroleum, shale oil, the U.S. tar sands, Canadian bitumen, refinery streams, and liquid fossil fuels. They are usually classified into two major groups: basic and nonbasic compounds. In nonbasic aromatic N-heterocyclic compounds, the lone pair of nitrogen atoms is part of the hetero-aromatic system and this lone electron pair is perpendicular to the plane of the molecule. Figure 9.1 shows some representative nonbasic aromatic N-heterocyclic compounds abundantly present in petroleum, refinery streams, commercial gasoline, and in diesel and jet fuels.

Contrary to nonbasic aromatic N-heterocyclic compounds, in basic aromatic N-heterocyclic compounds the lone electron pair of nitrogen atoms is located in the plane of the molecule. Figure 9.2 shows some typical aromatic N-heterocyclic compounds that show basic character and are present in petroleum, refinery streams, and liquid fossil fuels.

Basic aromatic N-heterocyclic compounds, such as substituted quinolines (QUINs), can constitute up to 50% of the total amount of nitrogen organic compounds present in gasoline and kerosene. On the other hand, in the intermediate streams of petroleum refinery such as refinery oils, for example, light cycle oil (LCO), more than 90% of the total amount of nitrogen-containing organic compounds is present as nonbasic compounds (Singh et al. 2011) including alkyl-substituted indoles (INDs) and curbazoles (CBZs).

The aromatic N-heterocyclic compounds present in petroleum distillates, airline jet fuels, commercial ground transportation fuels, and commercial heating oils can cause instability in fuels (Sobkowiak et al. 2009) due to the formation of deposits (Stalick et al. 2000), changes in the color of fuels, and an unpleasant odor. Some substituted INDs such as skatole (3-methylindole) present in liquid fossil fuels (Mushrush et al. 1990)

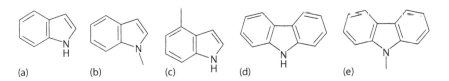

FIGURE 9.1 The representative nonbasic aromatic N-heterocyclic compounds: (a) indole, (b) 1-methylindole, (c) 4-methylindole, (d) carbazole, and (e) 9-methylcarbazole.

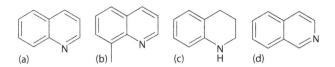

FIGURE 9.2 The representative basic aromatic N-heterocyclic compounds: (a) quinoline, (b) 8-methylquinoline, (c) 1,2,3,4-tetrahydroquinoline, and (d) isoquinoline.

and refinery streams have a very unpleasant odor. Large-ring aromatic N-heterocyclic compounds such as acridine and alkyl-substituted acridines are known to be carcinogenic (Satoh et al. 1997). In addition, there are data on the carcinogenicity of CBZ (Tsuda et al. 1982) and substituted CBZs, and on their acute aquatic toxicity. Combustion of liquid fossil fuels and by-products of the petroleum refinery process causes air pollution due to the formation of toxic nitrogen oxides NO_x, and acid rains.

In the industrial-scale commercial processes for the removal of nitrogen-containing organic compounds from refinery streams, catalytic hydrodenitrogenation (HDN) is conducted at an elevated temperature and pressure (Rodríguez and Ancheyta 2004), simultaneously with catalytic hydrodesulfurization (HDS). Aromatic N-heterocyclic compounds are known to adversely affect the performance of HDS (Murti et al. 2003). Specifically, methyl-substituted CBZs such as 1,8-dimethylcarbazole are known to be highly refractory in both HDN and HDS (Shin et al. 2001). Yet another difficulty accompanying catalytic HDS and HDN of refinery streams is undesired catalytic hydrogenation of aromatic hydrocarbons. This side reaction of hydrogenation causes an undesired decrease in the octane number (ON) of the obtained gasoline and an unnecessary consumption of hydrogen.

9.2 ADSORPTIVE DENITROGENATION OF LIQUID FOSSIL FUELS

A number of alternative processes for denitrogenation of liquid fossil fuels and refinery streams have been proposed. Particularly, selective adsorption of aromatic N-heterocyclic compounds from liquid fossil fuels in the presence of large amounts of aromatic hydrocarbons is a promising alternative to HDN. Since the early 2000s, adsorptive removal of aromatic N-heterocyclic compounds from liquid fossil fuels has been extensively studied, mostly on zeolites (Hernandez-Maldonado and Yang 2004) and activated carbons (Almarri et al. 2009). Recently, adsorption has been extended toward more "exotic" sorbents such as ion-exchange resins (Xie et al. 2010) and ionic liquids (ILs) (Huh et al. 2009).

Mesoporous metal–organic frameworks (MOFs) with their very high surface area and large pore sizes are suitable candidates for the adsorptive removal of aromatic N-heterocyclic compounds from liquid fossil fuels and refinery streams. Many mesoporous MOFs have moderate-to-high stability toward elevated temperatures and oxygen. This stability can be beneficial in applications based on adsorption that require sorbent regeneration by desorption in the nonaqueous phase under heating. On the other hand, hydrolysis, which is a rather common problem in adsorption-based applications of MOFs (e.g., Bezverkhyy et al. 2014), is not relevant in the case of adsorption/desorption from the nonaqueous phase in liquid fossil fuels and streams of petroleum refineries. Therefore, selective, multicycle adsorption/desorption on mesoporous MOFs can be utilized in the adsorptive denitrogenation of liquid fossil fuels, at ambient or moderately elevated temperatures.

In order to conduct a systematic search for combinations of "the MOF/substrate" that would yield the desired performance in the adsorptive denitrogenation of liquid fuels, one needs to have a solid understanding of mechanisms of adsorption and the atomic-level structure of adsorption active site(s) in the mesoporous MOFs of interest. In this chapter, specific attention is paid to the mechanisms of adsorption and desorption of aromatic N-heterocyclic compounds and the experimental techniques employed to learn about these adsorption mechanisms. Furthermore, along with N-heterocyclic compounds, liquid fossil fuels also contain rather large amounts of aromatic hydrocarbons. These aromatic hydrocarbons may compete with N-heterocyclic compounds in the adsorption on MOFs in solution.

9.3 ADSORPTION OF AROMATIC N-HETEROCYCLIC COMPOUNDS FROM FOSSIL FUELS ON MIL-101

Table 9.1 shows published data on adsorption of aromatic N-heterocyclic compounds from hydrocarbon solvents, namely, from model liquid fuels and commercial liquid fossil fuels, on mesoporous MOFs from the MIL-101 group, in the early period of research.

Adsorption of IND, CARB, and several methyl-substituted INDs and CBZs was studied on MIL-101(Cr) at ambient temperature (Nuzhdin et al. 2010). The model liquid fossil fuel studied contained IND as the target compound at few hundred ppmw of total nitrogen content. The solvent was either commercial hydrotreated straight run gas oil (SRGO) with the total nitrogen content as low as 10.8 ppmw N before the addition of IND or pure isooctane. On the other hand, the SRGO-based model liquid fuel contained as much as 25.0 wt.% aromatic hydrocarbons. The adsorption capacity of MIL-101(Cr) toward IND decreased by only a few percent, when isooctane as the solvent was replaced with hydrotreated SRGO, which contains aromatic hydrocarbons. Therefore, adsorption of IND was considered to be selective in the presence of a high concentration of competing aromatic hydrocarbons present in the SRGO. The adsorption mechanism of competing aromatic hydrocarbons in the SRGO was proposed as the π–π stacking interactions with the 1,4-benzenedicarboxylic acid (BDC) linker in MIL-101(Cr) (Nuzhdin et al. 2010); however, no experimental proof has been provided. One can note that in addition to π–π stacking interactions, nonspecific dispersive interactions could contribute to adsorption. The high adsorption capacity

TABLE 9.1
Adsorption of Aromatic N-Heterocyclic Compounds from Hydrocarbon Solvents on Pure MIL-101 in 2010–2013

MOF	Activated?	Adsorbate	Solvent	Adsorbed Amount	Bonding	Reference
MIL-101(Cr)(F)	150°C, air	IND	Isooctane	~18 mg N/g at C_e = 90 ppmw N	Via Cr(III)?	Nuzhdin et al. (2010)
MIL-101(Cr)(F)	Above	IND	SRGO, 25 wt.% aromatics	~18 mg N/g at C_e = 170 ppmw N	Via Cr(III)?	Nuzhdin et al. (2010)
MIL-101(Cr)(F)	Above	CBZ	Above	~6 mg N/g at C_e = 30 ppmw N	Via Cr(III)?	Nuzhdin et al. (2010)
MIL-101(Cr)(F)	423 K, air	IND; 2M; 1,2DMI NMCBZ	80:20 v/v n-heptane/toluene; 20:80 v/v n-heptane/toluene	IND > 2MI > 1,2DMI ~ NMC at C_0 = 0.15 M	?	Maes et al. (2011)
MIL-101(Cr)(F)	423 K, vacuum	PYRID; PYRR; QUIN; IND	n-octane; 75:25 v/v n-octane/p-xylene	PYRR Q_0 = 48.82 mg N/g < IND Q_0 = 49.44 mg N/g < PYRID Q_0 = 62.50 mg N/g < QUIN Q_0 = 64.69 mg N/g at 303 K	Lewis acid/base with Cr(III)	Wang et al. (2013b)

and selectivity of MIL-101(Cr) to aromatic N-heterocyclic compounds has been explained by their coordination interactions with the Cr(III) coordinatively unsaturated site (CUS). This conclusion seems questionable, since no thermal activation of the as-synthesized MIL-101(Cr) was reported to have been performed (Nuzhdin et al. 2010).

Adsorptive denitrogenation of a model liquid fuel with MIL-101(Cr) was studied by batch adsorption experiments at 303, 313, and 323 K (Wang et al. 2013b). Before adsorption, MIL-101(Cr) had been activated by drying in vacuum overnight at 423 K in order to generate the CUS. The MIL-101(Cr) sorbent was characterized by x-ray diffraction (XRD), nitrogen adsorption, Fourier transform infrared (FTIR) spectroscopy, and the Hammett acid–base indicator method. The model fuel contained pyridine (PYRID), pyrrole (PYRR), QUIN, or IND in n-octane as the solvent. The adsorption kinetics followed the pseudo-second-order kinetic rate law, and the thermodynamics of adsorption was fitted with the Langmuir equation (Wang et al. 2013b). The activated MIL-101(Cr) with the CUS Lewis acid sites and a high specific surface area adsorbed the following amounts at 303 K, in mg N/g sorbent: 48.82 for PYRR, 49.44 for IND, 62.50 for PYRID, and 64.99 for QUIN. This clearly indicates that acid–base interactions of the MIL-101(Cr) with aromatic N-heterocyclic compounds were mainly responsible for the observed adsorbed amounts. When the adsorption temperature was increased from 303 to 323 K, the adsorbed amount of each N-heterocyclic compound decreased, which indicates physisorption and, therefore, the possibility for easy regeneration of the sorbent (Wang et al. 2013b). Indeed, regeneration of the "spent" MIL-101(Cr) sorbent was performed by just washing it with ethanol at room temperature, followed by thermal reactivation of the MOF at 423 K. The structure of the regenerated MIL-101(Cr) was checked by XRD and FTIR. Although the crystallinity of regenerated MIL-101(Cr) decreased somewhat, the adsorption capacity for QUIN decreased by only a few percent over the five adsorption/desorption cycles (Wang et al. 2013b).

Table 9.2 shows the data on the adsorption/desorption of aromatic N-heterocyclic compounds on the MIL-101 MOFs published in more recent years.

The adsorption of QUIN versus IND was studied on MIL-101(Cr) by experiment and density functional theory (DFT) calculations (Wu et al. 2014). The adsorption isotherms of both basic QUIN and nonbasic IND follow the Langmuir isotherm. The adsorption capacity for QUIN was found to be higher than that for IND, which is consistent with the calculated higher bonding energy, specifically BE(QUIN) = −61.31 kJ/mol compared to BE(IND) = −38.33 kJ/mol. The DFT calculations also suggest that the adsorption of QUIN is dominated by the acid/base interactions of the lone pair of electrons on the N atom with the CUS of Cr(III) in MIL-101(Cr). On the other hand, in the adsorption of nonbasic IND, the main mechanism was hydrogen bonding. The addition of tetrahydrofuran (THF) to the octane solvent caused a strong decrease in the adsorbed amounts of aromatic N-heterocyclic compounds. The bonding energy of the THF to the Cr(III) CUS on MIL-101(Cr) was calculated to be rather high: BE(THF) = −60.62 kJ/mol. This explains experimentally observed, strongly competing effects of THF for adsorption sites in MIL-101(Cr).

In the other paper (Bhadra et al. 2015), the mechanism of adsorption was investigated by changing the polarity of the solvent in the adsorptive denitrogenation

TABLE 9.2
Adsorption of Aromatic N-Heterocyclic Compounds from Hydrocarbon Solvents on Pure MIL-101 Since 2014

MOF	Activation Procedure	Adsorbate	Solvent	Sorbent Characterization	Adsorbed Amount	Bonding	Reference
MIL-101(Cr)(F)	180°C, in vacuum	IND (indole)	Octane; octane/THF	Nitrogen sorption, XRD, SEM	$Q_{max} = 25.61$ mg/g	Hydrogen bonding	Wu et al. (2014)
MIL-101(Cr)(F)	180°C, in vacuum	QUIN (quinoline)	Octane; octane/THF	Above	$Q_{max} = 32.57$ mg/g	Acid–base interactions with Cr(III) CUS	Wu et al. (2014)
MIL-101(Cr)	100°C, in vacuum	Pyrrole	n-octane	—	$Q_0 = 100$ mg/g	Polar interactions	Bhadra et al. (2015)

experiment. Fluorine-free MIL-101(Cr) was synthesized first. After hydrothermal synthesis, MIL-101(Cr) was treated with water solution of NH_4F followed by drying in air and then in vacuum at 100°C overnight. For nonbasic PYRR, the adsorbed amount was the highest when the dielectric polarity of the solvent was the lowest (n-octane). Therefore, adsorption is believed to occur via polar interactions.

Postsynthetic modification (PSM) has been suggested to improve the selectivity and capacity of adsorption on mesoporous MOFs. The effects of the PSM of MIL-101 on the adsorptive properties are illustrated in Table 9.3.

The study of adsorption of quinoline (QUIN) versus IND on MIL-101(Cr) versus MIL-101(Cr) impregnated with phosphotungstic acid (PWA), the PWA/MIL-101(Cr), was reported (Ahmed et al. 2013b). To prepare PWA/MIL-101(Cr), MIL-101(Cr) was mixed with water, PWA was added and stirred for 3 h, and the mixture dried. Model liquid fuels contained 300 ppm QUIN or IND, and the solvent was the mixture of 75 vol.% n-octane and 25 vol.% p-xylene. The total surface area of PWA/MIL-101(Cr) at 2447 m^2/g was smaller than that of MIL-101(Cr) at 3028 m^2/g. Nevertheless, for basic QUIN, the adsorption capacity was 229 mg/g on MIL-101(Cr) versus 274 mg/g on PWA/MIL-101(Cr). Therefore, there was an increase of the adsorbed amount, apparently due to the interactions of the adsorbate with supported PWA acid. On the other hand, for nonbasic IND, the adsorption capacity was 162 mg/g on MIL-101(Cr) versus 152 mg/g for PWA/MIL-101(Cr). Therefore, the deposition of PWA did not increase the adsorption of the nonbasic aromatic N-heterocyclic compound, as expected. The adsorption capacity of regenerated PWA/MIL-101(Cr) after washing with ethanol was somewhat lower compared to that of regenerated MIL-101(Cr).

The same group investigated the adsorption of QUIN versus IND on the composite sorbent GO/MIL-101(Cr) formed by MIL-101(Cr) with graphene oxide (GO) (Ahmed et al. 2013c). The porosity of MIL-101(Cr) increased in the presence of small amounts of GO at <0.5 wt.%, but it decreased with higher content of GO. Before adsorption, the MOF sorbents were dried in vacuum at 150°C for 12 h to generate the CUS. The model fuels contained QUIN (300–1200 ppm) or IND (300–1200 ppm). For basic compound QUIN, the adsorption capacity was 481 mg/g on MIL-101(Cr) versus as much as 549 mg/g on GO/MIL-101(Cr). The higher adsorbed amount on GO/MIL-101(Cr) was explained by the higher total surface area at 3858 m^2/g, as compared to MIL-101(Cr) at 3155 m^2/g. For IND, the adsorption capacity was found at 244 mg/g on MIL-101(Cr) versus 319 mg/g on GO/MIL-101(Cr). These reported adsorbed amounts are much higher than those reported earlier by the same group (Ahmed et al. 2013b) for adsorption of IND and QUIN on MIL-101(Cr). The difference is apparently due to the higher concentration of the aromatic N-heterocyclic compounds (300–1200 ppm) used in the study where higher adsorbed amounts were reported, as expected. The deposition of GO caused an increase of adsorbed amount of nonbasic IND as well as basic QUIN. No spectroscopic study of the interactions between the adsorbate and the MOF with supported GO was conducted. A tentative explanation for the increased adsorption capacity was provided as being due to increased porosity of the composite sorbent GO/MIL-101(Cr) versus MIL-101(Cr). As concluded by the authors (Ahmed et al. 2013c), GO/MIL-101(Cr) showed the highest adsorption capacity for QUIN among

TABLE 9.3

Adsorption of Aromatic N-Heterocyclic Compounds from Hydrocarbon Solvents on Pure versus Modified MIL-101

MOF	Activation Procedure	Postmodified?	Adsorbate	Solvent	Adsorbed Amount	Bonding	Reference
MIL-101(Cr)(F)	150°C, in vacuum for 12 h	No	QUIN; IND	75:25 v/v; n-octane/p-xylene	QUIN $Q_0 = 229$ mg/g; IND $Q_0 = 162$ mg/g	Cr(III)?	Ahmed et al. (2013b)
PWA/MIL-101(Cr)(F)	Above	With acid PWA	QUIN; IND	Above	QUIN $Q_0 = 274$ mg/g; IND $Q_0 = 152$ mg/g	Brønsted acid/base, QUIN with PWA	Ahmed et al. (2013b)
MIL-101(Cr)(F)	150°C, in vacuum for 12 h	No	QUIN; IND	72:25 v/v; n-octane/p-xylene	QUIN $Q_0 = 481$ mg/g; IND $Q_0 = 244$ mg/g	—	Ahmed et al. (2013c)
GO/MIL-101(Cr)(F)	Above	No	QUIN; IND	Above	QUIN $Q_0 = 549$ mg/g; IND $Q_0 = 319$ mg/g	$\pi-\pi$ interactions with GO?	Ahmed et al. (2013c)

all sorbents reported by 2013. The "spent" GO/MIL-101(Cr) sorbent was washed with ethanol; it can be reused with a moderate decrease of the adsorbed amount at ca. 15% after four regeneration cycles. Different sorbent regeneration solvents have been tested (Figure 9.3) where QUI denotes quinoline and BT is benzothiophene; the best regeneration solvent was reported to be ethanol.

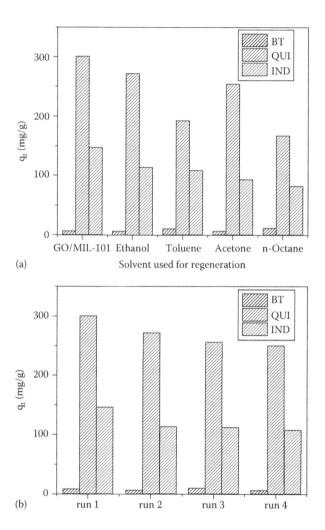

FIGURE 9.3   (a) Effect of solvents applied in the regeneration of the used 0.25% GO/MIL-101 on the adsorptive removal of BT, QUI, and IND. (b) Effect of regeneration cycles on the performances of adsorptive removal of BT, QUI, and IND over regenerated 0.25% GO/MIL-101 by washing with ethanol (run 1 represents the fresh adsorbent). The hydrocarbon solutions containing 600 ppm BT, 300 ppm QUI, and 300 ppm IND. (Reprinted with permission from Ahmed, I., Khan, N.A., and Jhung, S.H., Graphite oxide/metal-organic framework (MIL-101): Remarkable performance in the adsorptive denitrogenation of model fuels, *Inorg. Chem.*, 52(24), 14155–14161. Copyright 2013 American Chemical Society.)

Surface functionalization of MIL-101(Cr) by PSM with triflic anhydride and sulfuric acid in nitromethane as the solvent was first reported in 2011 (Goesten et al. 2011). The obtained sulfated S-MIL-101(Cr) has been characterized by solid-state nuclear magnetic resonance (NMR), x-ray absorption near-edge structure (XANES), and FTIR spectroscopy. S-MIL-101(Cr) showed Brønsted acidity due to the sulfoxy acid groups attached to up to 50% of the terephthalate linkers, and it showed catalytic activity in the esterification of n-butanol with acetic acid. Subsequently, this synthesis was utilized (Wang et al. 2013a) to obtain S–MIL-101(Cr) for an adsorptive denitrogenation of model liquid fossil fuels. The S–MIL-101(Cr) sorbent contained 0.21–0.42 mmol/g sulfonic acid groups and retained the structure of MIL-101(Cr), but the specific surface areas and pore diameters decreased. S–MIL-101(Cr) showed an increase in the adsorption capacity by 12.2% for basic QUIN and 6.3% for nonbasic IND versus MIL-101(Cr) in n-octane as the solvent. Regeneration of the "spent" sorbent was conducted via washing with ethanol, and the adsorption capacity did not decrease after three adsorption/desorption cycles (Wang et al. 2013a).

9.4   ADSORPTION OF AROMATIC N-HETEROCYCLIC COMPOUNDS FROM FOSSIL FUELS ON MIL-100

Table 9.4 shows published data on adsorption on another major kind of mesoporous MOF, MIL-100, in its chemically nonmodified form, for the aromatic N-heterocyclic compounds from the liquid model and commercial fossil fuels.

Adsorption of several substituted INDs and CBZs was studied at 298 K on isostructural MIL-100(Fe), MIL-100(Cr), and MIL-100(Al) (Maes et al. 2011). The adsorption on these MIL-100 materials has also been compared with adsorption on MIL-101(Cr)(F). Before adsorption, MIL-100(Cr) and MIL-101(Cr) were thermally activated at 423 K to generate the CUS of Cr(III), while MIL-100(Al) was heated at 373 K. On the other hand, MIL-100(Fe) was washed with hot water and ethanol after synthesis to remove the excess of the 1,3,5-benzenetricarboxylic (BTC) linker but was not thermally activated. Therefore, one cannot expect to observe the effects of the presence of Fe(III) CUS on adsorption. Liquid model fuels contained IND, 2 methylindole, 1,2-dimethylindole, or N-methylcarbazole at a constant initial concentration of 0.15 M. The solvent was either 80 vol.% n-heptane/20 vol.% toluene or 20 vol.% n-heptane/80 vol.% toluene; hence, the effect of competitive adsorption of aromatic hydrocarbon toluene could be estimated. Indeed, significant competitive adsorption of toluene was found, which lowered the adsorption capacity for all aromatic N-heterocyclic compounds studied (Maes et al. 2011). For example, for IND on MIL-100(Fe), the adsorption capacity was 36 wt.% (of the total amount of IND initially present) in the solvent containing 80 vol.% n-heptane, but it was only 16 wt.% in the solvent containing 80 vol.% toluene. Nevertheless, the adsorption capacities were considered to be potentially usable; the highest maximum adsorption uptake was 40 wt.% for MIL-100(Al) in the model fuel containing 80 vol.% n-heptane. The proposed mechanism of adsorption was the interactions between the CUS metal sites in the MOFs and the molecules of aromatic N-heterocyclic adsorbate. This apparently does not apply to MIL-100(Al), therefore, its adsorption mechanism must have been different.

TABLE 9.4
Adsorption of Aromatic N-Heterocyclic Compounds from Hydrocarbon Solvents on Pure MIL-100, 2011–2013

MOF	Activated?	Adsorbate	Solvent	Characterization	Adsorbed Amount	Reference
MIL-100(Cr)(F)	423 K	IND, 2MI, 1,2DMI, NMCBZ	n-heptane + toluene	Microcalorimetry, Mössbauer spectrum	IND > 2MI > 1,2DMI > NMC	Maes et al. (2011)
MIL-100(Fe)(F)	No	IND, 2MI, 1,2DMI, NMCBZ	Above	Above	IND > 2MI > 1,2DMI > NMC	Maes et al. (2011)
MIL-100(Al)	373 K	IND, 2MI, 1,2DMI, NMCBZ	Above	Above	IND > 2MI > 1,2DMI > NMC	Maes et al. (2011)
MIL-100(V)	423 K air/vacuum	IND; 1,2DMI	n-heptane	N_2 adsorption, BET, FTIR, microcalorimetry	370 molecules/cell; 290 molecules/cell	Van de Voorde et al. (2013)
MIL-100(Fe)(F)	423 K	IND; 1,2DMI	n-heptane	Above	670 molecules/cell; 300 molecules/cell	Van de Voorde et al. (2013)
MIL-100(Al)	423 K	IND; 1,2DMI	n-heptane	Above	670 molecules/cell; 240 molecules/cell	Van de Voorde et al. (2013)
MIL-100(Cr)(F)	423 K	IND; 1,2DMI	n-heptane	Above	560 molecules/cell; 340 molecules/cell	Van de Voorde et al. (2013)

In the follow-up article (Van de Voorde et al. 2013), the same group investigated the adsorption of IND and 1,2-dimethylindole (1,2DMI) on isostructural MIL-100 with Al(III), Cr(III), Fe(III), and vanadyl VO(III) sites. Sample characterization was performed by measuring adsorption isotherms, microcalorimetry, and FTIR spectroscopic study. Prior to the adsorption, MIL-100(Fe), MIL-100(Cr), and MIL-100(Al) were thermally activated overnight at 423 K in air in an oven. On the other hand, MIL-100(V) was thermally activated in vacuum ($<10^{-5}$ Pa) at 423 K for 12 h. The adsorption isotherms were measured at 298 K in n-heptane as the solvent. The type of metal in MIL-100 was found to strongly affect the adsorption capacity and the integral adsorption enthalpies. Depending on the MIL-100 used, the values of initial integral adsorption enthalpies ΔH_{ads} were higher for either IND or 1,2DMI, which implied different adsorption mechanisms for different metals in MIL-100. For MIL-100(Fe) and MIL-100(Al), the number of adsorbed IND molecules was significantly larger than the number of available CUSs; therefore, an adsorption site was not clearly identified. For the FTIR spectroscopic study, an activated MIL-100 was immersed into the solution of IND in n-heptane and n-heptane was evaporated by heating the sample at 323 K. Thus, evaporation of the solvent in ambient humid air, rather than in vacuum, causes oxidation and/or hydrolysis of the organic adsorbate. Then, the sample was pressed into the pellet, outgassed in vacuum, and the transmission FTIR spectra were measured. It was assumed that physisorbed IND would evaporate, while chemisorbed IND would remain in the sample. Since MIL-100 MOFs strongly absorbs at >1600 cm^{-1}, the FTIR spectroscopic study was concentrated mainly on the analysis of vibrations at 2900–3500 cm^{-1}. Chemisorbed IND was detected via the vibrations at 3250 and 2900 cm^{-1} that were found to be affected by hydrogen bonding to different degrees for the different metals in MIL-100. For MIL-100(Al), no hydrogen bonds were formed, so it was speculated that only coordination bonds were present. The nature of this coordination was not explained. On the other hand, for MIL-100(Fe) and MIL-100(V) saturated with IND, N-H vibrations between 3500 and 3200 cm^{-1} were detected, which indicated hydrogen bonding of adsorbed IND to the MOFs (Figure 9.4, Table 9.4).

The adsorption of QUIN versus IND was studied on MIL-100(Fe) experimentally and based on DFT calculations (Wu et al. 2014) (Table 9.5); as in the case of MIL-101(Cr), the adsorbed amount for QUIN was higher than for IND, apparently due to the basicity of the former.

The mesoporous MOF Basolite F300 is similar to MIL-100(Fe) in its chemical composition and structural parameters. Specifically, the iron content is 25 versus 21 wt.%, the carbon content is 32 versus 29 wt.%, and the mesopore size is 21.7 Å versus 25 and 29 Å for Basolite F300 and MIL-100(Fe), respectively (Dhakshinamoorthy et al. 2012). Recently, we reported the mechanistic study of adsorption of IND and naphthalene (NAP) on Basolite F300 by two complementary spectroscopic methods and DFT calculations (Dai et al. 2014). The excitation wavelength–dependent fluorescence spectra, the near-UV/VIS diffuse reflectance spectra (DRS) and the DFT calculations suggest that IND forms an adsorption complex with Basolite F300 MOF. In this complex, the IND molecule is electronically bound to the Fe(III) CUS and dispersively stabilized by the side interactions with benzene rings of the BTC linker (see Figure 9.5). Coordination bonds between an adsorbed

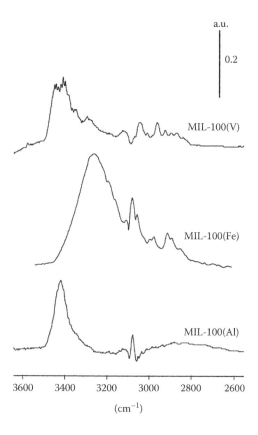

FIGURE 9.4 Spectra of adsorbed species persisting on MIL-100 after outgassing at 423 K, subtracted from spectra at 373 K. (Reprinted with permission from Van de Voorde, B., Boulhout, M., Vermoortele, F., Horcajada, P., Cunha, D., Lee, J.S., Chang, J.S. et al., N/S-heterocyclic contaminant removal from fuels by the mesoporous metal-organic framework MIL-100: The role of the metal ion, *J. Am. Chem. Soc.*, 135(26), 9849–9856. Copyright 2013 American Chemical Society.)

IND and Fe(III) in the activated Basolite F300 MOF were also detected by geometry optimization using DFT calculations. We also reported the electronic and dispersive contributions to the binding energy (BE) versus total BE for IND adsorbed on Basolite F300. In addition, electronic spectra of the adsorption complex of IND with Basolite F300 MOF as calculated by the time-dependent DFT (TD-DFT) method suggest the formation of the ligand-to-metal charge transfer (LMCT) complex. The computed UV-Vis spectra are in agreement with the experimental UV-Vis DRS spectra, which show the new absorption bands at 460–660 nm due to the adsorption complex. The kinetics of formation of the adsorption complex by the chemical reaction of pure IND with activated Basolite F300 was found to follow the zero-order rate law (Dai et al. 2014).

Table 9.6 shows the role of PSM of MIL-100 on the adsorption of N-heterocyclic compounds in nonaqueous solutions.

TABLE 9.5
Adsorption of Aromatic N-Heterocyclic Compounds from Hydrocarbon Solvents on MIL-100 and Similar MOFs Since 2014

MOF	Activation Procedure	Adsorbate	Solvent	Characterization	Adsorbed Amount	Bonding	Reference
MIL–100(Fe)	180°C in vacuum	QUIN	Octane, octane/THF	N_2 adsorption, XRD, SEM	$Q_{max} = 34.48$ mg/g	Acid–base interactions with Cr(III) CUS	Wu et al. (2014)
MIL–100(Fe)	180°C in vacuum	IND	Octane, octane/THF	Above	$Q_{max} = 13.08$ mg/g	Hydrogen bonding	Wu et al. (2014)
Basolite F300	100°C in vacuum	IND	None	UV-Vis DRS, PL, DFT, TD-DFT	—	LMCT with Fe(III)	Dai et al. (2014)

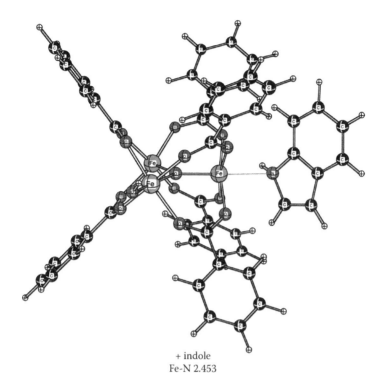

+ indole
Fe-N 2.453

FIGURE 9.5 Optimized structure (B3LYP/LANL2DZ) of complex formed by cluster $Fe_3O(OH)L_6$ with indole. Large gray atoms, Fe; black atoms, C; small white atoms, H; red atoms, O; blue atom, N (color figure online). (With kind permission from Springer Science + Business Media: *J. Porous Mater.*, Adsorption of naphthalene and indole on F300 MOF in liquid phase by the complementary spectroscopic, kinetic and DFT studies, 21(5), 2014, 709–727, Dai, J., McKee, M.L., and Samokhvalov, A.)

Jhung's group reported the PSM of MIL-100(Cr) by grafting the surface acidic or basic functional groups, and studies of adsorption of aromatic N-heterocyclic compounds on such postsynthetically modified MOFs (Ahmed et al. 2013a). Specifically, the basic sorbent ED–MIL-100(Cr) was prepared by grafting ethylenediamine (ED) $H_2N-CH_2-CH_2-NH_2$ onto the Cr(III) CUS in MIL-100(Cr), as has been reported earlier for MIL-101(Cr) (Hwang et al. 2008). The acidic counterpart AMSA–MIL-100(Cr) was synthesized by grafting aminomethanesulfonic acid (AMSA) $H_2N-CH_2-SO_3H$ onto the Cr(III) CUS of MIL-100(Cr) (Ahmed et al. 2013a). Grafting was confirmed by elemental analysis and FTIR spectra. The grafted MOFs demonstrated the typical XRD pattern of MIL-100(Cr), and the porosity decreased in the order MIL-100(Cr) > ED–MIL-100(Cr) > AMSA–MIL-100(Cr). The model fuel contained QUIN or IND dissolved in the mixture of 75 vol.% n-octane and 25 vol.% p-xylene. For basic QUIN, the adsorption capacity was 342 mg/g on MIL-100(Cr), 119 mg/g on ED–MIL-100(Cr), and 323 mg/g on AMSA–MIL-100(Cr) (Ahmed et al. 2013a). The data indicated, surprisingly, that there was no increase in the adsorbed amount of

TABLE 9.6

Adsorption of Aromatic N-Heterocyclic Compounds from Hydrocarbon Solvents on Postsynthetically Modified MIL-100

MOF	Activation	Postmodified?	Adsorbate	Solvent	Adsorbed Amount	Bonding	Reference
MIL-100(Cr)(F)	150°C for 12 h	No	QUIN; IND	75:25 v/v; n-octane/p-xylene	QUIN Q_0 = 342 mg/g; IND Q_0 = 103 mg/g	—	Ahmed et al. (2013a)
ED-MIL-100(Cr)(F)	150°C for 12 h	Base ED	QUIN; IND	Above	QUIN Q_0 = 119 mg/g; IND Q_0 = 94 mg/g	Brønsted acid/base	Ahmed et al. (2013a)
AMSA-MIL-100(Cr)(F)	150°C for 12 h	Acid AMSA	QUIN; IND	Above	QUIN Q_0 = 323 mg/g; IND Q_0 = 97 mg/g	—	Ahmed et al. (2013a)
MIL-100(Cr)(F)	100°C	No	QUIN; IND	75:25 v/v; n-octane/p-xylene	QUIN Q_0 = 420 mg/g; IND Q_0 = 149 mg/g	—	Ahmed and Jhung (2014)
CuCl/MIL-100(Cr)(F)	100°C	CuCl	IND; QUIN	Above	QUIN Q_0 = 457 mg/g; IND Q_0 = 171 mg/g	π-complex?	Ahmed and Jhung (2014)

QUIN on the acidic sorbent AMSA–MIL-100(Cr) compared to MIL-100(Cr). These relatively high adsorbed amounts were obtained from the solvent containing 25 vol.% aromatic hydrocarbon p-xylene. Therefore, one can conclude that basic QUIN has a rather low capacity on the functionalized basic ED–MIL-100(Cr), while it has much higher adsorption capacities on both the nonmodified and acid-functionalized AMSA–MIL-100(Cr). The presumed acid–base interactions have not been studied by spectroscopic or calorimetric methods. For the nonbasic IND, the adsorption capacity was 103 mg/g on MIL-100(Cr), 94 mg/g on ED–MIL-100(Cr), and 97 mg/g on AMSA–MIL-100(Cr). Therefore, the adsorption of nonbasic IND is not very sensitive to the surface acidity or basicity in the MOF. Overall, acid-grafted AMSA–MIL-100(Cr) has been selected as the best sorbent; this sorbent can be used several times for adsorption of QUIN and IND after regeneration by washing with acetone. The conclusion was that for effective adsorptive removal of basic aromatic N-heterocyclic compounds, acid–base interactions with functional groups of the MOFs can be used. On the other hand, for adsorptive removal of nonbasic aromatic N-heterocyclic compounds, some other mechanism of adsorption, for example, π–π interaction, π complexation with the CUS, or hydrogen bonding is needed (Ahmed et al. 2013a).

The Cu(I) cation is a weak Lewis acid, and this property can be used in the adsorption of basic N-aromatic compounds in solution. MIL-100(Cr) postmodified by the deposition of CuCl yielded the novel composite sorbent CuCl/MIL-100(Cr) (Ahmed and Jhung 2014). CuCl/MIL-100(Cr) has been compared to MIL-100(Cr) in the adsorptive removal of basic QUIN versus nonbasic IND from a model fuel with 75% n-octane/25% p-xylene. At the initial concentrations of QUIN or IND of 400 ppm, the adsorption capacities for MIL-100(Cr) were 420 mg/g for QUIN and 149 mg/g for IND. For the composite sorbent CuCl/MIL-100(Cr), the adsorption capacities were found to be higher: 457 mg/g for QUIN and 171 mg/g for IND, despite the total surface area of CuCl/MIL-100(Cr) being lower than that of MIL-100(Cr). Therefore, it can be concluded (Ahmed and Jhung 2014) that the increased adsorption capacities of the composite CuCl/MIL-100(Cr) were due to the formation of π-complexes between the N-aromatic adsorbate and Cu^+ sites.

An analysis of the available literature leads us to the following conclusions. In the majority of reported studies on adsorption of aromatic N-heterocyclic compounds, mesoporous MOFs are expected to contain the CUS of the transition metal. There are only few papers that report adsorptive denitrogenation on the mesoporous MOF with an Al(III) site (Maes et al. 2011, Van de Voorde et al. 2013). The competing adsorption of aromatic hydrocarbons is not very severe in the adsorptive denitrogenation on mesoporous MOFs, as compared to adsorption of aromatic sulfur compounds (Samokhvalov 2015). Indeed, the first paper on adsorptive denitrogenation with mesoporous MOFs (Nuzhdin et al. 2010) reported rather high adsorbed amounts of IND and substituted carbazoles from gas oils that contained high concentrations of aromatic hydrocarbons. In the mechanistic studies of adsorption of aromatic N-heterocyclic compounds on mesoporous MOFs, several spectroscopic studies of the adsorption complexes have been reported that evidence the formation of either hydrogen bonds (Van de Voorde et al. 2013) or coordination bonds (Dai et al. 2014) between the aromatic N-heterocyclic adsorbate and the CUS in the mesoporous MOFs as sorbents.

REFERENCES

Ahmed, I., Z. Hasan, N. A. Khan, and S. H. Jhung. 2013a. Adsorptive denitrogenation of model fuels with porous metal-organic frameworks (MOFs): Effect of acidity and basicity of MOFs. *Applied Catalysis B: Environmental* 129:123–129.

Ahmed, I. and S. H. Jhung. 2014. Adsorptive denitrogenation of model fuel with CuCl-loaded metal-organic frameworks (MOFs). *Chemical Engineering Journal* 251:35–42.

Ahmed, I., N. A. Khan, Z. Hasan, and S. H. Jhung. 2013b. Adsorptive denitrogenation of model fuels with porous metal-organic framework (MOF) MIL-101 impregnated with phosphotungstic acid: Effect of acid site inclusion. *Journal of Hazardous Materials* 250:37–44.

Ahmed, I., N. A. Khan, and S. H. Jhung. 2013c. Graphite oxide/metal-organic framework (MIL-101): Remarkable performance in the adsorptive denitrogenation of model fuels. *Inorganic Chemistry* 52(24):14155–14161.

Almarri, M., X. Ma, and C. Song. 2009. Role of surface oxygen-containing functional groups in liquid-phase adsorption of nitrogen compounds on carbon-based adsorbents. *Energy & Fuels* 23:3940–3947.

Bezverkhyy, I., G. Ortiz, G. Chaplais, C. Marichal, G. Weber, and J. P. Bellat. 2014. MIL-53(Al) under reflux in water: Formation of gamma-AlO(OH) shell and H_2BDC molecules intercalated into the pores. *Microporous and Mesoporous Materials* 183:156–161.

Bhadra, B. N., K. H. Cho, N. A. Khan, D. Y. Hong, and S. H. Jhung. 2015. Liquid-phase adsorption of aromatics over a metal-organic framework and activated carbon: Effects of hydrophobicity/hydrophilicity of adsorbents and solvent polarity. *Journal of Physical Chemistry C* 119(47):26620–26627.

Dai, J., M. L. McKee, and A. Samokhvalov. 2014. Adsorption of naphthalene and indole on F300 MOF in liquid phase by the complementary spectroscopic, kinetic and DFT studies. *Journal of Porous Materials* 21(5):709–727.

Dhakshinamoorthy, A., M. Alvaro, P. Horcajada, E. Gibson, M. Vishnuvarthan, A. Vimont, J. M. Greneche, C. Serre, M. Daturi, and H. Garcia. 2012. Comparison of porous iron trimesates Basolite F300 and MIL-100(Fe) as heterogeneous catalysts for Lewis acid and oxidation reactions: Roles of structural defects and stability. *ACS Catalysis* 2(10):2060–2065.

Goesten, M. G., J. Juan-Alcañiz, E. V. Ramos-Fernandez, K. B. Sai Sankar Gupta, E. Stavitski, H. van Bekkum, J. Gascon, and F. Kapteijn. 2011. Sulfation of metal–organic frameworks: Opportunities for acid catalysis and proton conductivity. *Journal of Catalysis* 281(1):177–187.

Hernandez-Maldonado, A. J. and R. T. Yang. 2004. Denitrogenation of transportation fuels by zeolites at ambient temperature and pressure. *Angewandte Chemie—International Edition* 43(8):1004–1006.

Huh, E. S., A. Zazybin, J. Palgunadi, S. Ahn, J. Hong, H. S. Kim, M. Cheong, and B. S. Ahn. 2009. Zn-containing ionic liquids for the extractive denitrogenation of a model oil: A mechanistic consideration. *Energy & Fuels* 23(6):3032–3038.

Hwang, Y. K., D.-Y. Hong, J.-S. Chang, S. H. Jhung, Y.-K. Seo, J. Kim, A. Vimont, M. Daturi, C. Serre, and G. Férey. 2008. Amine grafting on coordinatively unsaturated metal centers of MOFs: Consequences for catalysis and metal encapsulation. *Angewandte Chemie—International Edition* 47(22):4144–4148.

Maes, M., M. Trekels, M. Boulhout, S. Schouteden, F. Vermoortele, L. Alaerts, D. Heurtaux et al. 2011. Selective removal of N-heterocyclic aromatic contaminants from fuels by Lewis acidic metal-organic frameworks. *Angewandte Chemie—International Edition* 50(18):4210–4214.

Murti, S. D. S., H. Yang, K. H. Choi, Y. Korai, and I. Mochida. 2003. Influences of nitrogen species on the hydrodesulfurization reactivity of a gas oil over sulfide catalysts of variable activity. *Applied Catalysis A: General* 252(2):331–346.

Mushrush, G. W., E. J. Beal, R. N. Hazlett, and D. R. Hardy. 1990. Interactive effects of 2,5-dimethylpyrrole, 3-methylindole, and tert-butyl hydroperoxide in a shale-derived diesel fuel. *Energy & Fuels* 4(1):15–19.

Nuzhdin, A. L., K. A. Kovalenko, D. N. Dybtsev, and G. A. Bukhtiyarova. 2010. Removal of nitrogen compounds from liquid hydrocarbon streams by selective sorption on metal-organic framework MIL-101. *Mendeleev Communications* 20(1):57–58.

Rodríguez, M. A. and J. Ancheyta. 2004. Modeling of hydrodesulfurization (HDS), hydrodenitrogenation (HDN), and the hydrogenation of aromatics (HDA) in a vacuum gas oil hydrotreater. *Energy & Fuels* 18(3):789–794.

Samokhvalov, A. 2015. Adsorption on mesoporous metal–organic frameworks in solution: Aromatic and heterocyclic compounds. *Chemistry—A European Journal* 21(47):16726–16742.

Satoh, K., H. Sakagami, T. Kadofuku, T. Kurihara, and N. Motohashi. 1997. Radical intensity and carcinogenic activity of benz[c]acridines. *Anticancer Research* 17(5A):3553–3557.

Shin, S., H. Yang, K. Sakanishi, I. Mochida, D. A. Grudoski, and J. H. Shinn. 2001. Inhibition and deactivation in staged hydrodenitrogenation and hydrodesulfurization of medium cycle oil over NiMoS/Al$_2$O$_3$ catalyst. *Applied Catalysis A: General* 205(1–2):101–108.

Singh, D., A. Chopra, M. B. Patel, and A. S. Sarpal. 2011. A comparative evaluation of nitrogen compounds in petroleum distillates. *Chromatographia* 74(1–2):121–126.

Sobkowiak, M., J. A. Griffith, B. Wang, and B. Beaver. 2009. Insight into the mechanisms of middle distillate fuel oxidative degradation. Part 1: On the role of phenol, indole, and carbazole derivatives in the thermal oxidative stability of Fischer-Tropsch/petroleum jet fuel blends. *Energy & Fuels* 23:2041–2046.

Stalick, W. M., S. Faour, R. V. Honeychuck, B. Beal, B. R. Hardy, and G. W. Mushrush. 2000. Fuel instability studies: The reaction of 3-methyl indole with aryl sulfonic acids. *Petroleum Science and Technology* 18(3–4):287–304.

Tsuda, H., A. Hagiwara, M. Shibata, M. Ohshima, and N. Ito. 1982. Carcinogenic effect of carbazole in the liver of (C57BL/6N × C3H/HeN)F1 mice. *Journal of the National Cancer Institute* 69(6):1383–1389.

Van de Voorde, B., M. Boulhout, F. Vermoortele, P. Horcajada, D. Cunha, J. S. Lee, J. S. Chang et al. 2013. N/S-heterocyclic contaminant removal from fuels by the mesoporous metal-organic framework MIL-100: The role of the metal ion. *Journal of the American Chemical Society* 135(26):9849–9856.

Wang, Z. Y., G. Li, and Z. G. Sun. 2013a. Denitrogenation through adsorption to sulfonated metal-organic frameworks. *Acta Physico–Chimica Sinica* 29(11):2422–2428.

Wang, Z. Y., Z. G. Sun, L. H. Kong, and G. Li. 2013b. Adsorptive removal of nitrogen-containing compounds from fuel by metal-organic frameworks. *Journal of Energy Chemistry* 22(6):869–875.

Wu, Y., J. Xiao, L. M. Wu, M. Chen, H. X. Xi, Z. Li, and H. H. Wang. 2014. Adsorptive denitrogenation of fuel over metal organic frameworks: Effect of N-types and adsorption mechanisms. *Journal of Physical Chemistry C* 118(39):22533–22543.

Xie, L.-L., A. Favre-Reguillon, X.-X. Wang, X. Fu, and M. Lemaire. 2010. Selective adsorption of neutral nitrogen compounds from fuel using ion-exchange resins. *Journal of Chemical & Engineering Data* 55(11):4849–4853.

10 Adsorption of Aromatic Sulfur Compounds from Liquid Fuels

10.1 AROMATIC SULFUR COMPOUNDS IN LIQUID FOSSIL FUELS

Sulfur is a chemical element present in crude petroleum between approximately 100 and 33,000 ppm in its elemental as well as its chemically bound forms (Samokhvalov and Tatarchuk 2010). Aromatic sulfur compounds such as thiophene, benzothiophene (BT), dibenzothiophene (DBT) (Figure 10.1), and their methyl-substituted counterparts are abundantly present in virtually all fossils and liquid fossil fuels: coal, petroleum, shale oil, U.S. tar sands, and Canadian bitumen (Speight 2000), in the streams of petroleum refineries (Ito and van Veen 2006), in commercial gasoline and diesel (Samokhvalov and Tatarchuk 2010), and in jet fuels and heating oils.

When liquid fossil fuels are burned, toxic and corrosive sulfur dioxide (SO_2) and/or sulfate particulate matter (PM) are emitted in air, which contributes to smog and acid rains. In the mid-2000s, ultralow-sulfur (ULS) gasoline (30 ppmw total sulfur) and ultralow-sulfur diesel (ULSD, 15 ppmw total sulfur) standards were implemented in the United States. Similar standards were enforced in the EU (Samokhvalov 2012). Still, certain commercial liquid fuels in the United States contain very high concentrations of aromatic sulfur compounds. For instance, commercial U.S. heating oils can contain up to 10,000 ppmw of total sulfur, which is present mostly as fused-ring aromatic sulfur compounds. The latter compounds remain in commercially available liquid fuel after standard catalytic hydrodesulfurization (HDS) at the petroleum refinery. In addition, military liquid logistic fuels in the United States can contain up to 5,000 ppmw of total sulfur (Samokhvalov 2012), which is present therein as fused-ring aromatic sulfur compounds; see Table 10.1.

In developed countries, there was recently a successful transition to environment-friendly ULS ground transportation fuels with total sulfur at or below 30 ppmw (Table 10.2).

Many developing countries are transitioning from high-sulfur fuels (many hundreds ppmw S) to low-sulfur and ULS liquid ground transportation fuels. The total content of chemically bound sulfur in aviation fuels has historically been very high (Table 10.3); this situation arose because the content of sulfur in the airline jet fuels is not regulated.

The content of total sulfur in liquid hydrocarbon fuels for civil propeller airplanes and jets exploited in the United States remains very high. The total content of sulfur in U.S. military logistic fuels is also very high. Worldwide distributed operations of the U.S. Navy, Air Force, and Army need to rely on *locally available* (i.e., with high

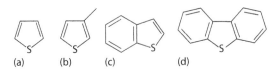

FIGURE 10.1   Representative aromatic sulfur compounds. (a) Thiophene; (b) 3-methylthiophene (3-MT); (c) benzothiophene (BT); (d) dibenzothiophene (DBT).

TABLE 10.1
Total Concentration of Sulfur in the U.S. Military Liquid Logistic Fuels

	Jet Fuel for Navy Aircraft	Jet Fuel for the Air Force	Jet Fuel for the Air Force	Diesel Fuel for Navy Ships	Diesel Fuel for Ground Combat Vehicles
Kind of liquid fuel	JP-5, F-44	JP-7	JP-8, F-34, F-35	F-76	—
Total content of sulfur in fuel, ppmw	70-4,000	1,000	140-3,000	<5,000	Exempt from EPA standards

TABLE 10.2
Total Concentration of Sulfur in Civil Gasoline

Kind of Fuel	Ground Transportation Gasoline				Aviation Gasoline	
Country/region	U.S.	EU	Japan	Russia	U.S.	UK
Commercial fuel	ULS gasoline			Class 3 motor fuel	Avgas 100LL	100LL
Total content of sulfur, ppmw	30	<10	<10	<150	500	500

TABLE 10.3
Total Concentration of Sulfur in Civil Jet Fuels

Country	United States	United Kingdom	CIS[a]	Canada
Name of fuel	Jet A	Jet A-1	TS-1	Jet B
Total content of sulfur, ppmw	3,000	3,000	2,500	4,000

[a] C.I.S. = Commonwealth of Independent States (the former USSR).

concentrations of sulfur compounds in many cases) liquid fuels, rather than on high-quality ULS fuels to be delivered from the United States. This is the reason behind the interest of the U.S. military in research on emerging portable technologies for purification of liquid fuels that can be effectively deployed on-site and on demand to produce ultralow or "zero-sulfur" fuels for fuel cell (FC) applications from the available liquid logistic fossil fuels (e.g., Tatarchuk et al. 2008, Yang et al. 2008, Dhage et al. 2011).

10.2 METHODS OF DESULFURIZATION OF LIQUID FOSSIL FUELS

Catalytic HDS is the standard industrial-scale method for deep desulfurization (hundreds of ppmw total sulfur) and ultradeep desulfurization (low tens of ppmw total sulfur) of liquid fossil fuels. HDS requires high pressures up to 50 atm, temperatures of 350°C–400°C, and corrosion-resistant trickle bed reactors. Various modifications of HDS for desulfurization of refinery streams have been comprehensively reviewed (e.g., Babich and Moulijn 2003, Stanislaus et al. 2010). Industrial catalytic HDS of petroleum feedstock is not very effective in the selective removal of large-ring alkyl-substituted aromatic sulfur compounds. One of the major reaction pathways in catalytic desulfurization is catalytic hydrogenation. Olefins and aromatic compounds are hydrogenated together with aromatic sulfur compounds, which causes lowering of the octane number (ON) of the obtained commercial liquid fuel. Another major disadvantage of HDS is that its major gaseous reaction product is highly corrosive and very toxic hydrogen sulfide.

Several alternative methods of desulfurization of liquid fossil fuels have been reported and reviewed: adsorptive desulfurization (Hernandez-Maldonaldo and Yang 2004, Samokhvalov and Tatarchuk 2010), oxidative catalytic desulfurization (ODS) (Campos-Martin et al. 2010, Ismagilov et al. 2011), desulfurization by chemical oxidation (Ito and van Veen 2006), photochemical desulfurization (Samokhvalov 2012), photocatalytic desulfurization (Samokhvalov 2011), and bio-desulfurization (Boniek et al. 2014). Amongst these methods, nonreactive adsorption of aromatic sulfur compounds present in liquid fossil fuels is one of the most well-studied alternatives to catalytic HDS. Nonreactive adsorption occurs when the molecule of an aromatic sulfur compound binds reversibly to the active site in the sorbent. Therefore, there are no other products of chemical reaction than the surface "adsorption complex" (e.g., Zhu et al. 2013, Dai et al. 2014, Demir et al. 2014). The adsorption complex is generated due to chemical bonds formed between the adsorbate molecule and the adsorption site. The adsorption complex is expected to be destroyed during the following regeneration of the "spent" sorbent, and an active form of the adsorption site in the sorbent and desorbed compounds are thus formed. Mechanistic studies of reactive adsorption (e.g., Ryzhikov et al. 2008) and nonreactive adsorption of aromatic sulfur compounds in the liquid phase (Samokhvalov et al. 2011) have been conducted and reviewed (Samokhvalov and Tatarchuk 2010). Nonreactive desulfurization by adsorption is the most promising route toward commercialization. The well-known "conventional" sorbents for desulfurization by nonreactive adsorption are zeolites with supported cations of transition metals (e.g., Hernandez-Maldonaldo and Yang 2004). In recent years, some advanced sorbents such as ionic liquids (ILs) have been extensively investigated for selective desulfurization of liquid fossil fuels and streams of petroleum refineries (Martinez-Palou and Luque 2014). Mesoporous metal–organic frameworks (MOFs) have certain advantages as emerging sorbents for the removal of aromatic sulfur compounds from model and "real" liquid fossil fuels.

10.3 ADSORPTION OF AROMATIC SULFUR COMPOUNDS FROM LIQUID FOSSIL FUELS ON MIL-101

Table 10.4 provides information on adsorption and desorption of aromatic sulfur compounds from model and real fuels on pure MIL-101 that was published in the early period of these studies, that is, throughout 2010–2012.

Adsorption of DBT, 4-methyldibenzothiophene (4-MDBT), and 4,6-DMDBT has been studied for a model fuel on *as-synthesized* MIL-101(Cr) that was not thermally activated and thus contained no coordinatively unsaturated sites (CUSs) of the metal (Nuzhdin et al. 2010). The model fuel contained isooctane or commercial hydrotreated straight run gas oil (SRGO) as a solvent. This SRGO contained as much as 25.0 wt. % aromatic hydrocarbons. The adsorbed amount of DBT decreased by the factor ~2.5 when SRGO was used as the solvent instead of isooctane. Therefore, there was significant competitive adsorption of aromatic hydrocarbons on MIL-101(Cr). This competitive adsorption was tentatively explained by both aromatic sulfur compounds and aromatic hydrocarbons having the same adsorption mechanism, the π–π stacking interactions with the 1,4-benzenedicarboxylic acid (BDC) linker in MIL-101(Cr); however, no experimental proof was provided.

TABLE 10.4

Adsorption and Desorption of Aromatic Sulfur Compounds from Fuels on Pure MIL-101, 2010–2012

MOF	Activation	Adsorbate	Solvent	Adsorbed Amount	Reference
MIL-101(Cr)(F)	150°C in air overnight	DBT	Isooctane	10 mg S/g at C_e = 450 ppmw S	Nuzhdin et al. (2010)
MIL-101(Cr)(F)	Above	4 MDBT	Isooctane	7 mg S/g at C_e = 350 ppmw S	Nuzhdin et al. (2010)
MIL-101(Cr)(F)	Above	4,6-DMDBT	Isooctane	7 mg S/g at C_e = 350 ppmw S	Nuzhdin et al. (2010)
MIL-101(Cr)(F)	Above	DBT	SRGO	3 mg S/g at C_e = 380 ppmw S	Nuzhdin et al. (2010)
MIL-101(Cr)(F)	423 K, in air overnight	T	n-heptane/ toluene	3% at C_0 = 0.15 M T < BT < DBT	Maes et al. (2011)
MIL-101(Cr)(F)	Above	BT	n-heptane/ toluene	5% at C_0 = 0.15 M	Maes et al. (2011)
MIL-101(Cr)(F)	Above	DBT	n-heptane/ toluene	8% at C_0 = 0.15 M	Maes et al. (2011)
MIL-101(Cr)(F)	150°C overnight	3-MT; BT; DBT	n-octane	3-MT ~ 22 g S/g; BT ~ 32 mg S/g; DBT ~41 mg S/g	Zhang et al. (2012)

TABLE 10.5

Adsorption and Desorption of Aromatic Sulfur Compounds from Fuels on Pure MIL-101, 2012–2016

MOF	Activation	Adsorbate	Solvent	Adsorbed Amount	Reference
MIL-101(Cr)(F)	180°C, vacuum	BT; DBT; 4,6-DMDBT	Octane	BT Q = 25.15 μmol S/g DBT Q = 78.17 μmol S/g 4,6-DMDBT Q = 67.03 μmol S/g	Wu et al. (2014a)
MIL-101(Cr)(F)	180°C, vacuum	DBT	Octane/THF/ naphthalene	—	Wu et al. (2014b)
MIL-101(Cr)(F)	180°C, vacuum	4,6-DMDBT	Octane/THF/ naphthalene	—	Wu et al. (2014b)
MIL-101(Cr)(F)	DMF + CH_2Cl_2	Thiophene; BT; 4,6-DMDBT	Isooctane	Thiophene 0.12 mmol S/g; BT 0.21 mmol S/g; 4,6-DMDBT 0.22 mmol S/g	Li et al. (2015)
MIL-101(Cr)		Thiophene	n-octane	$Q_0 = 45$ mg/g	Bhadra et al. (2015)

The adsorption of T, BT, and DBT from solution in the same solvent mixture (n-heptane and toluene) was reported on MIL-101(Cr) (Maes et al. 2011). The adsorbed amounts of aromatic sulfur compounds on MIL-101(Cr) were found to be very low. This has been explained by the strong competition of the aromatic solvent toluene with the target aromatic sulfur compounds for the adsorption sites. The mechanism of adsorption of aromatic sulfur compounds and aromatic hydrocarbons on MIL-101 was not studied by experiment.

The adsorption of DBT and 4,6-DMDBT from octane has been studied on three MOFs, and the adsorption capacity was found to be lower on mesoporous MIL-101(Cr) than on microporous MOF Cu–BTC (Wu et al. 2014a). The mechanism of adsorption of DBT and 4,6-DMDBT was studied by DFT calculations by the same group. DBT and 4,6-DMDBT were found to be adsorbed onto the CUS of Cr^{3+} at the angles 39.6° and 12.7°, respectively (Wu et al. 2014b).

This suggests that both the conjugated π-system and a lone pair of electrons on the sulfur atom contribute to adsorption (Wu et al. 2014b). It has also been concluded that the methyl groups in 4,6-DMDBT exhibit both the electron donor and sterical hindrance effects, and the latter decreases the extent of the former. The binding energy (BE) of 4,6-DMDBT at −24.02 kJ/mol was found to be less exothermic than that of DBT at −30.52 kJ/mol. This is consistent with the lower adsorption selectivity of 4,6-DMDBT as found in the experiments with solvents containing, in addition to octane, aromatic hydrocarbon naphthalene and oxygenate cosolvent tetrahydrofuran (THF) (Wu et al. 2014b).

The adsorption of thiophene, BT, and 4,6-DMDBT was studied on MIL-101(Cr) (Li et al. 2015). The mechanism of adsorption of thiophene was studied by Fourier transform infrared (FTIR) spectroscopy and thermogravimetric analysis (TGA). Pure thiophene has a strong characteristic band at 701 cm^{-1} due to the S–C stretching vibration. After adsorption on MIL-101(Cr), this band was found at 710 cm^{-1}, which corresponds to a blue shift of 9 cm^{-1}. This was a moderate change compared to the blue shift of 15 cm^{-1} observed with thiophene adsorbed on microporous MIL-53(Cr). The TGA of thiophene adsorbed on MIL-101(Cr) revealed one desorption peak, compared to the two peaks for MIL-53(Cr). In addition, the adsorbed amount on MIL-101(Cr) was found to be much less than on MIL-53(Cr), so microporous MOF was found to be more advantageous for the adsorption of the rather small thiophene molecule.

The effect of polarity of the solvent on the adsorbed amount of thiophene on MIL-101(Cr) has been studied (Bhadra et al. 2015). The adsorbed amount was found to be higher when the dielectric polarity of the solvent was lower, and the adsorption was explained by the polar interactions between the adsorbates with active sites in the MOF.

10.4 ADSORPTION OF AROMATIC SULFUR COMPOUNDS FROM LIQUID FOSSIL FUELS ON MODIFIED MIL-101

The effects of postsynthetic modification (PSM) of MIL-101 on the adsorption of aromatic sulfur compounds are provided in Tables 10.6 and 10.7. MIL-101(Cr) was impregnated with phosphotungstic acid (PWA) to form the composite material PWA/MIL-101(Cr) with acidic surface groups (Ahmed et al. 2013b). Adsorption of BT was studied on PWA/MIL-101(Cr) and MIL-101(Cr). For BT, only 30 mg/g on MIL-101(Cr) was adsorbed, in contrast to 229 mg/g sorbent for quinoline (QUIN) and 162 mg/g sorbent for indole (IND). For PWA/MIL-101(Cr) with acidic surface groups, 26 mg BT/g sorbent was adsorbed. Therefore, the creation of acidic surface sites did not increase the adsorptive performance of MIL-101(Cr) toward the aromatic sulfur compound BT.

The composites of graphene oxide (GO) with MIL-101(Cr) were synthesized by the introduction of GO into the mixture of precursors for MIL-101(Cr) (Ahmed et al. 2013c). Structural characterization by x-ray diffraction (XRD) and the Brunauer–Emmett–Teller (BET) method showed that the composites retained the structure of MIL-101(Cr), and they had a bimodal pore size distribution, as expected of MIL-101, at ca. 15 and 20 Å. The porosity of the composites increased in the presence of <0.5 % GO, but further decreased at the higher loaded amounts. The adsorption of BT from the model fuel (the mixture of 75 % n-octane and 25 % xylene) resulted in the maximum adsorbed amount of BT at 28.2 mg/g sorbent. The adsorption of BT from the model fuel in the presence of QUIN and/or IND nitrogen aromatic compounds with about the same size of the fused ring was found to be negligibly small. This indicates the preferred adsorption of aromatic N-heterocyclic compounds versus the respective aromatic sulfur compounds. When only BT was present in the model fuel, the adsorbed amount of BT showed a linear correlation with the surface

TABLE 10.6

Adsorption and Desorption of Aromatic Sulfur Compounds from Fuels on Pure versus Modified MIL-101, 2013

MOF	Activation	Postmodification	Adsorbate	Solvent	Adsorbed Amount	Reference
MIL-101(Cr)(F)	150°C, 12 h.	No	BT	75:25 v/v n-octane/p-xylene	$Q_0 = 30$ mg/g	Ahmed et al. (2013b)
PWA/MIL-101(Cr)(F)	No	Acid PWA	BT	Above	$Q_0 = 26$ mg/g	Ahmed et al. (2013b)
MIL-101(Cr)(F)	150°C in vacuum, 12 h.	No	BT	Above	$Q_0 = 18.3$ mg/g	Ahmed et al. (2013c)
GO/MIL-101(Cr)(F)	150°C in vacuum, 12 h.	GO	BT	Above	$Q_0 = 28.2$ mg /g	Ahmed et al. (2013c)
MIL-101(Cr)(F)	150°C in vacuum	No	DBT	n-octane	$Q_0 = 28.94$ mg/g	Jia et al. (2013)
PTA/MIL-101(Cr)(F)	Above	PSM with PTA	DBT	Above	—	Jia et al. (2013)
PTA@MIL-101(Cr)(F)	Above	One-pot synthesis	DBT	Above	$Q_0 = 107.63$ mg/g	Jia et al. (2013)

TABLE 10.7

Adsorption and Desorption of Aromatic Sulfur Compounds from Fuels on Pure versus Modified MIL-101, Since 2014

MOF	Activation Procedure	Postmodification	Adsorbate	Solvent	Adsorbed Amount	Reference
MIL-101(Cr)(F)	100°C in vacuum overnight	No	BT; DBT	n-octane	BT: $Q_0 = 51$ mg/g; DBT: ~35 mg/g at $C_0 = 1000$ ppm	Khan et al. (2014)
IL/MIL-101(Cr)	Dehydrated	Acidic ionic liquid (IL)	BT; DBT	n-octane	BT: $Q_0 = 87$ mg/g; DBT: ~55 mg/g at $C_0 = 1000$ ppm	Khan et al. (2014)
MIL-101(Cr)–SO$_3$H	105°C in vacuum for 12 h	No	Thiophene; BT; DBT	n-octane	Thiophene: 5.47 g S/kg; BT: 7.95 g S/kg; DBT: 2.14 g S/kg	Huang et al. (2015)
MIL-101(Cr)–SO$_3$Ag	Above	Ion exchange with AgNO$_3$	Thiophene; BT; DBT	n-octane	Thiophene: 20.5 g S/kg; BT: 28.8 g S/kg; DBT: 31.0 g S/kg	Huang et al. (2015)

area of the composites. This implies mostly π–π and nonspecific interactions with the aromatic linker BDC (which creates the majority of the total surface area), rather than interactions with the relatively small CUS of metal in the MOF. However, the role of Cr(III) CUS was not ruled out by the authors; thus, the mechanism of selective adsorption of aromatic N-heterocyclic compounds versus aromatic sulfur compounds on the GO/MIL-101(Cr) composites remains unclear.

Composite sorbents containing phosphotungstic acid denoted PTA and MIL-101(Cr) were prepared by "one-pot" synthesis from precursors and through PSM of already prepared MIL-101(Cr) (Jia et al. 2013). The samples were characterized by nitrogen adsorption, transmission electron microscopy (TEM), XRD, and FTIR spectroscopy. The introduction of PTA to MIL-101(Cr) during the synthesis caused a 10 % increase in the adsorption capacity versus MIL-101(Cr). On the other hand, the PTA/MIL-101(Cr) prepared by PSM of MIL-101(Cr) demonstrated lower adsorption capacity. The theoretical maximum adsorption capacity Q_0 for PTA@MIL-101(Cr) prepared by one-pot method was 136.5 mg S/g, which is about 4.2 times higher than that for MIL-101(Cr). Furthermore, PTA@MIL-101(Cr) and MIL-101(Cr) remained effective in the adsorption of DBT in the presence of competing aromatic hydrocarbons benzene, p-xylene, and naphthalene. After adsorption, the sorbents were regenerated in methanol under stirring and dried. The adsorption capacities for MIL-101(Cr) and PTA@MIL-101(Cr) decreased to 9.1 mg S/g and 10.0 mg S/g, respectively. According to inductively coupled plasma (ICP) atomic emission spectroscopic analysis of the regeneration solvent, no tungsten was leached during sorbent regeneration (Jia et al. 2013).

Adsorptive desulfurization of liquid fuels with ILs was reported (e.g., Martinez-Palou and Luque 2014). Acidic IL prepared by the interaction of 1-butyl-3-methylimidazolium chloride with $AlCl_3$ was supported on MIL-101(Cr) to form the composite sorbent IL/MIL-101(Cr) (Khan et al. 2014). IL/MIL-101(Cr) was studied to adsorb BT and DBT from n-octane. The adsorption kinetics followed the pseudo-second-order rate law. The adsorption capacity of IL/MIL-101(Cr) for BT was 87 mg/g sorbent, compared to 51 mg/g for pure MIL-101(Cr). The increased adsorbed amount for IL/MIL-101(Cr) versus MIL-101(Cr) was tentatively explained by the acid–base interactions between the supported acidic IL and BT. The type of acidity involved (Lewis or Brønsted type) was not reported, and no experiment to determine acid–base interactions was conducted. The presence of toluene in the model fuels (10 or 25 vol. %) caused a decrease in the adsorption capacity of both MIL-101(Cr) and IL/MIL-101(Cr), due to competitive adsorption (Khan et al. 2014).

The cationic form of silver Ag+ has been frequently utilized as an active site in sorbents for desulfurization of liquid fossil fuels (Samokhvalov et al. 2010b, Samokhvalov and Tatarchuk 2010). It has been shown that the π-system of the molecule of the aromatic sulfur compound interacts as a soft Lewis base with Ag+ present on the surface of the sorbent, which acts as a soft Lewis acid (Samokhvalov et al. 2010a). Recently, an interesting paper reported, for the first time, controlled modification of mesoporous MIL-101(Cr) with Ag+ for adsorptive desulfurization in solution (Huang et al. 2015). First, MIL-101(Cr)–SO_3H was synthesized from sodium salt of 2-sulfoterephthalic acid, $Cr(NO_3)_3$, and aqueous HF, and then dried and activated in vacuum at 105°C for 12 h. Then, the obtained MIL-101(Cr)–SO_3H was added to the solution of $AgNO_3$ in CH_3CN/H_2O at room temperature for 24 h for ion exchange

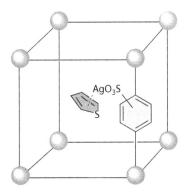

FIGURE 10.2 Metal-organic framework with –SO_3Ag for adsorptive desulfurization. (From Huang, M.H., Chang, G.G., Su, Y., Xing, H.B., Zhang, Z.G., Yang, Y.W., Ren, Q.L., Bao, Z.B., and Chen, B.L., A metal-organic framework with immobilized Ag(I) for highly efficient desulfurization of liquid fuels, *Chem. Commun.*, 51(61), 12205–12207, 2015. Reproduced by permission of The Royal Society of Chemistry.)

and washed. Finally, the obtained MIL-101(Cr)–SO_3Ag was dried in vacuum over-night. The adsorption isotherms for thiophene in octane indicated that MIL-101(Cr)–SO_3Ag adsorbs much higher amounts of thiophene than MIL-101(Cr)–SO_3H, despite the higher total surface area of the latter. Therefore, the adsorption mechanism was attributed to the interaction of thiophene with the Ag+ site in the MOF (Figure 10.2).

The "spent" sorbent was regenerated by washing with methanol or acetone at room temperature. No significant loss of adsorption capacity was detected after three cycles of "adsorption–regeneration" (Huang et al. 2015).

10.5 ADSORPTION OF AROMATIC SULFUR COMPOUNDS FROM LIQUID FOSSIL FUELS ON MIL-100

The MIL-100 family has also been investigated in the adsorption of aromatic sulfur compounds from liquid fuels. Table 10.8 shows the data on adsorption of aromatic sulfur compounds on chemically nonmodified MIL-100. The adsorption of aromatic sulfur compounds T, BT, and DBT was studied on MIL-100(Fe), MIL-100(Cr), and MIL-100(Al) (Maes et al. 2011).

The solvent was a mixture of n-heptane and toluene at the 80:20 v/v and 20:80 v/v ratios, in order to test the competing adsorption of aromatic hydrocarbons. In all cases, the adsorbed amounts of aromatic sulfur compounds were found to be very low and practically not useful for adsorption-based applications, due to strong adsorption of toluene on all the MOFs studied. It has been concluded that the CUSs of Fe(III), Cr(III), and Al(III), which are the "hard" Lewis acids, do not interact strongly with the sulfur atom (a "soft" Lewis base) in the molecule of aromatic sulfur compound (Maes et al. 2011).

Adsorption of 3-methylthiophene (3-MT), BT, and DBT from n-octane has been compared on MIL-100(Cr) and MIL-101(Cr) (Zhang et al. 2012). The model diesel

TABLE 10.8

Adsorption and Desorption of Aromatic Sulfur Compounds from Fuels on Pure MIL-100, 2011–2013

MOF	Activation Procedure	Adsorbate	Solvent	Adsorbed Amount	Reference
MIL-100(Cr)(F)	423 K	T; BT; DBT	n-heptane/ toluene	T 2%; BT 7 %; DBT 7% at $C_0 = 0.15M$	Maes et al. (2011)
MIL-100(Cr)(F)	150°C overnight	3-MT; BT; DBT	n-octane	DBT 30.7 mg S/g at $C_0 = 1000$ ppm; DBT > BT > 3-MT	Zhang et al. (2012)
MIL-100(Cr)(F)	423 K air	T	n-heptane	25% of total CUS	Van de Voorde et al. (2013)
MIL-100(Fe)(F)	423 K air	T	n-heptane	32% of total CUS	Van de Voorde et al. (2013)
MIL-100(Al)	423 K air	T	n-heptane	29% of total CUS	Van de Voorde et al. (2013)
MIL-100(V)	423 K air/ vacuum	T	n-heptane	15% of total CUS	Van de Voorde et al. (2013)

fuel in this study contained DBT, BT, or 3-MT dissolved in n-octane. The boiling point range of commercial diesel fuel is 260°C–360°C, while n-octane has its boiling point at 125°C. Thus, n-octane used as the solvent for model diesel in the study (Zhang et al. 2012) is not a representative solvent for diesel fuel. Instead, the longer-chain n-alkane should be used, so that the boiling point of the solvent is closer to the typical boiling range of commercial diesel fuel. The adsorption capacity for DBT followed the sequence MIL-101(Cr) > MIL-100(Cr). The adsorption capacity was found to decrease in the order DBT > BT > 3-MT, namely, the larger the molecule, the higher the adsorbed amount. The adsorption of DBT was explained as being due to an interplay of many factors: the framework structure, exposed Lewis acid sites/CUSs, suitable pore size, and pore shape. Thus, no single mechanism could be proposed for the observed adsorption. The "spent" MOF sorbent was regenerated by washing with methanol and drying at 150°C overnight. The regenerated MOF sorbent was reused in adsorption/desorption up to five times. FTIR spectroscopy was employed to compare "fresh" and regenerated MOF sorbents, and no substantial differences were found, indicating complete sorbent regeneration.

Adsorption of thiophene from the model fuel with n-heptane as the solvent at 298 K was studied on isostructural MIL-100(Fe), MIL-100(Cr), MIL-100(Al), and MIL-100(VO) (Van de Voorde et al. 2013). Depending on the MOF, 130–180 molecules were adsorbed per unit cell of the MOF at saturation. This adsorbed amount corresponds to adsorption on less than 1/3 of the total number of CUSs of metals. These results were qualitatively explained by the Pearson's hard and soft acids and bases (HSABs) concept: the sulfur atom in thiophene is the "soft" Lewis base that does not interact strongly with the "hard" Lewis acid sites—the CUSs of metals in the oxidation state +3 or higher. The role of other possible adsorption mechanisms, including the π–π interactions between the heteroaromatic ring of thiophene and the linkers, was not determined, and the authors concluded (Van de Voorde et al. 2013), "It seems that the factors determining the affinities of the materials for thiophene are not easy to unravel."

Table 10.9 shows the more recent (since 2014) efforts in the research on adsorptive desulfurization using nonmodified MIL-100. The study was published (Wu, Xiao, Wu, Xian, et al. 2014a) on adsorption of DBT and 4,6-DMDBT from octane on MIL-101(Cr) and MIL-100(Fe). The adsorption capacities were found to be rather low for both MOFs. For MIL-101(Cr), the adsorption capacity followed the order DBT > 4,6-DMDBT, while for MIL-100(Fe) the *opposite* order 4,6-DMDBT > DBT was observed. Apparently, both π-electrons and a lone pair of electrons on the S atom could have contributed to the interaction with MOF sorbents, as was discussed by the authors. One can note that these structure–property correlations are of limited value, since MIL-101(Cr) and MIL-100(Fe) feature *two* major structural differences: they have different CUS and different linkers.

The answer to the question of the relative contributions of these interactions can, in principle, be achieved by quantum chemical computations. Indeed, the authors have used density functional theory (DFT) calculations for computing BEs. For DBT, the BE on MIL-101(Cr) was −30.52 kJ/mol (Wu et al. 2014b), while on MIL-100(Fe) it was much less at −17.96 kJ/mol (Wu et al. 2014a). To investigate further, the electronic and dispersive components of the BE can be computed separately using the DFT method, for example, in our recent work (Dai et al. 2014).

MIL-100(Fe) was synthesized through a mechano-chemical route (ball milling) without using HF (Pilloni et al. 2015) and compared with the commercially available iron trimesate MOF Basolite F300 produced via the electrochemical route. The ball mill–prepared MIL-100(Fe) showed better crystallinity, good thermal stability, and higher surface area and pore volume than Basolite F300 MOF. The maximum

---

TABLE 10.9

Adsorption and Desorption of Aromatic Sulfur Compounds from Fuels on Pure MIL-100 Since 2014

MOF	Activation Procedure	Adsorbate	Solvent	Adsorbed Amount	Reference
MIL-100(Fe)(F)	180°C, in vacuum	BT	Octane	$Q = 17.52\ \mu mol\ S/g$	Wu et al. (2014a)
MIL-100(Fe)(F)	180°C, in vacuum	DBT	Octane	$Q = 45.37\ \mu mol\ S/g$	Wu et al. (2014a)
MIL-100(Fe)(F)	180°C, in vacuum	4,6-DMDBT	Octane	$Q = 61.48\ \mu mol\ S/g$	Wu et al. (2014a)
MIL-100(Fe), HF-free	378 K	4,6-DMDBT	n-heptane	$Q_{max} = 285\ \mu mol/g$	Pilloni et al. (2015)
Basolite F300	403 K	4,6-DMDBT	n-heptane	$Q_{max} = 165\ \mu mol/g$	Pilloni et al. (2015)

adsorption capacity of 4,6-DMDBT in the n-heptane solution for the ball-milled MOF was about twice as high compared to Basolite F300. To investigate the sites in MIL-100(Fe) responsible for adsorption of 4,6-DMDBT, ammonia adsorption calorimetry was used. The conclusion was that the sites that interact with ammonia with a differential heat of 90 kJ/mol are likely to be the same sites that adsorb 4,6-DMDBT, supposedly the CUS of Fe(III) (Pilloni et al. 2015).

10.6 ADSORPTION OF AROMATIC SULFUR COMPOUNDS FROM LIQUID FOSSIL FUELS ON MODIFIED MIL-100

PSM has been studied to improve the adsorption capabilities of MIL-100 MOFs (Table 10.10).

Contrary to Cu(II), which is a "hard" Lewis acid, Cu(I) is a "soft" Lewis acid. Therefore, Cu(I) would be a preferred adsorption site for aromatic sulfur compounds that are "soft" Lewis bases. The composite material $Cu_2O/MIL-100(Fe)$ was prepared by reducing the complex of Cu(II) in water in the presence of MIL-100(Fe) (Khan and Jhung 2012). The presence of the supported Cu(I) form was demonstrated by x-ray photoelectron spectroscopy (XPS). The adsorption capacity increased by up to 14 % by increasing the content of Cu(I) up to a Cu/Fe wt./wt. ratio of 0.07. The adsorption capacity for BT obtained from solution in n-octane was as high as 135–154 mg/g sorbent, depending on the amount of deposited Cu(I). This is a rather high adsorption capacity, compared to those obtained with other mesoporous MOFs, including MIL-100 (Maes et al. 2011). The mechanism of adsorption was speculated to be π-complexation between the BT and supported Cu(I), but no experimental proof was provided.

MIL-100(Cr) has been modified with surface Brønsted acidic or basic groups of organic compounds, and it was then used for the adsorption of BT versus nitrogen-heterocyclic compounds from model fuels (Ahmed et al. 2013a). Specifically, the surface modification of MIL-100(Cr) with the basic functional group was performed in the solution at room temperature by grafting ethylenediamine (ED) as was reported in a prior publication (Hwang et al. 2008) for MIL-101(Cr) (Figure 10.3). In surface-grafted MIL-100(Cr) used for the adsorption of BT (Ahmed 2013a), Pd was not used. The adsorbed amount of BT on basic ED–MIL-100(Cr) was 23 mg/g, which is lower than that on the nonmodified MIL-100(Cr) at 28 mg/g. Next, grafting of acidic surface sites was performed by adsorption of aminomethanesulfonic acid (AMSA) onto the CUS of activated MIL-100(Cr), to synthesize the acidic MOF AMSA–MIL-100(Cr).

The adsorption capacity of acidic AMSA–MIL-100(Cr) was as high as 323 mg/g for the aromatic N-heterocyclic compound (QUIN), but it was only 31 mg/g for the aromatic sulfur compound BT (Ahmed et al. 2013a). In addition, the adsorption capacity for BT on acidic AMSA–MIL-100(Cr) at 31 mg/g was just 10% higher than that on nonacidic MIL-100(Cr) at 28 mg/g, under the same conditions. The reported increase of adsorbed amount at 10% indicates that the surface modification of MIL-100(Cr) with a strong Brønsted acid did not allow any significant increase in the adsorbed amount of BT.

MIL-100(Cr)(F) was postsynthetically modified by impregnation with CuCl as prepared by chemical reduction of $CuCl_2$ in solution (Ahmed and Jhung 2014). The obtained composite CuCl/MIL-100(Cr) was tested in the adsorptive removal of BT

TABLE 10.10

Adsorption and Desorption of Aromatic Sulfur Compounds from Fuels on Pure versus Modified MIL-100

MOF	Activation Procedure	Postmodification	Adsorbate	Solvent	Adsorbed Amount	Reference
MIL-100(Fe)(F)	100°C, in vacuum	No	BT	n-octane	$Q_0 = 135$ mg BT/g	Khan and Jhung (2012)
Cu$_2$O/MIL-100(Fe)(F)	Above	Cu$_2$O	BT	n-octane	$Q_0 = 154$ mg BT/g	Khan and Jhung (2012)
MIL-100(Cr)(F)	150°C, in vacuum 12 h	No	BT	75:25 v/v n-octane/p-xylene	$Q_0 = 28$ mg/g	Ahmed et al. (2013a)
ED-MIL-100(Cr)(F)	Above	Base ED	BT	Above	$Q_0 = 23$ mg/g	Ahmed et al. (2013a)
AMSA-MIL-100(Cr)(F)	150°C, in vacuum 12 h	Acid AMSA	BT	Above	$Q_0 = 31$ mg/g	Ahmed et al. (2013a)
MIL-100(Cr)(F)	Above	No	BT	Above	~15 mg/g	Ahmed and Jhung (2014)
CuCl/MIL-100(Cr)(F)	Above	CuCl	BT	Above	~12 mg/g	Ahmed and Jhung (2014)

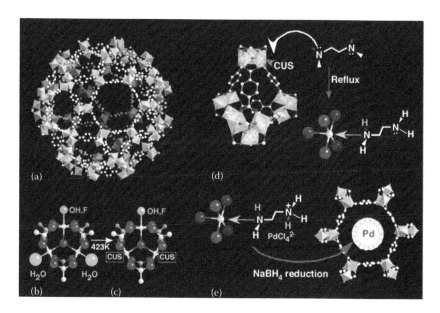

FIGURE 10.3 Site-selective functionalization of MIL-101 with unsaturated metal sites: (a) perspective view of the mesoporous cage of MIL-101 with hexagonal windows; (b,c) evolution of coordinatively unsaturated sites from chromium trimers in mesoporous cages of MIL-101 after vacuum treatment at 423 K for 12 h; (d) surface functionalization of the dehydrated MIL-101 through selective grafting of amine molecules (i.e., ethylenediamine) onto coordinatively unsaturated sites; (e) selective encapsulation of noble metals in the amine-grafted MIL-101 via a three-step process (see text). Chromium atoms/octahedra, yellow; carbon atoms, pale gray; oxygen atoms, red. (From Hwang, Y.K., Hong, D.-Y., Chang, J.-S., Jhung, S.H., Seo, Y.-K., Kim, J., Vimont, A., Daturi, M., Serre, C., and Férey, G.: Amine grafting on coordinatively unsaturated metal centers of MOFs: Consequences for catalysis and metal encapsulation. *Angew. Chem. Int. Ed.* 2008. 47(22). 4144–4148. Copyright Wiley-VCH Verlag GmbH & Co. KGaA. Reproduced with permission.)

from a model liquid fossil fuel containing 75%/25% n-octane/p-xylene. At the initial concentration of BT in solution of 800 ppm, the highest adsorbed amount was ca. 15 mg BT/g sorbent. The adsorbed amount was lower for CuCl/MIL-100(Cr) than for MIL-100(Cr), which correlates with the total surface areas: 1310 m²/g for CuCl/MIL-100(Cr) versus 1510 m²/g for MIL-100(Cr). Therefore, the beneficial effects of Lewis acid Cu(I) on the adsorption of BT from solution were absent, which can be explained by the significant competing adsorption of p-xylene, the major component of the solvent of the model fuel.

10.7 ADSORPTION OF AROMATIC SULFUR COMPOUNDS ON MESOPOROUS MOFs OTHER THAN MIL-100 AND MIL-101

There are only few reports on adsorption of aromatic sulfur compounds in solution on mesoporous MOFs other than MIL-101 and MIL-100. The new mesoporous MOF built from Cu(II) and the H_6L linker (Figure 10.4) and denoted (1) with formula

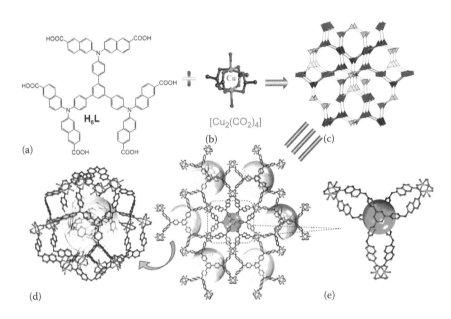

FIGURE 10.4 (a) The structure of H_6L. (b) Ball-and-stick representation of a paddle-wheel $[Cu_2(CO_2)_4]$ cluster. (c) Space-filling and ball-and-stick representations of the 3D net of 1. (d,e) Ball-and-stick views of two kinds of cages in 1. (From Li, S.L., Lan, Y.Q., Sakurai, H., and Xu, Q.: Unusual regenerable porous metal-organic framework based on a new triple helical molecular necklace for separating organosulfur compounds. *Chemistry—A European Journal*. 2012. 18(51). 16302–16309. Copyright Wiley-VCH Verlag GmbH & Co. KGaA. Reproduced with permission.)

$[Cu(L)_{1/3}(H_2O)]\cdot 8DMA$, where DMA = dimethylacetamide has been synthesized, characterized, activated, and tested for the adsorptive removal of BT and DBT from isooctane at room temperature (Li et al. 2012). The design of this new MOF has apparently been inspired by the studies of adsorptive desulfurization on microporous HKUST-1 (e.g., Cychosz et al. 2008).

HKUST-1 contains the "paddle-wheel" Cu_2O_8 dimers with the Cu(II) CUS and has a trimodal pore size distribution with the three cages. In HKUST-1, these are (1) a large cuboctahedral cage I 13 Å in diameter with coordinatively unsaturated Cu(II)-Cu(II) sites, (2) a large cuboctahedral cage II 11 Å in diameter, and (3) a small octahedral cage 4–5 Å in diameter (Senkovska et al. 2012). In the reported new MOF (Li et al. 2012) (Figure 10.4), the four carboxylate groups of different L_6 linkers form a paddle-wheel group $Cu_2(CO_2)_4$. The main structural feature of this new MOF is a mesocage 20.5 Å in size, and a microcage of 6.8 Å; therefore, this MOF can be considered a mesoporous analog of HKUST-1. The adsorption capacity of aromatic sulfur compounds was reported in the rather seldom used units of mmol adsorbate per cm^3 of free pore volume: 2.16 $mmol/cm^3$ for BT and 2.51 $mmol/cm^3$ for DBT, at the initial concentration of the aromatic sulfur compounds corresponding to 1500 ppmw of sulfur (Li et al. 2012). The regeneration of the "spent" sorbent was conducted by washing with isooctane for 24 h. The BET total surface area, the Langmuir surface area, and the pore volume for a regenerated MOF sorbent changed

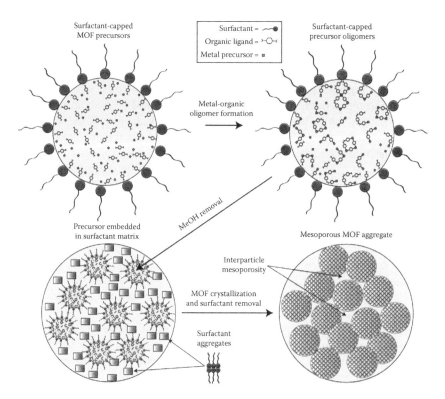

FIGURE 10.5 Schematic representation of mesoporous MIL-125 formation. (Reprinted with permission from McNamara, N.D. and Hicks, J.C., Chelating agent-free, vapor-assisted crystallization method to synthesize hierarchical microporous/mesoporous MIL-125 (Ti), *ACS Appl. Mater. Interfaces*, 7(9), 5338–5346. Copyright 2015 American Chemical Society.)

slightly. Regenerated MOF was used for adsorptive desulfurization again, and it demonstrated about the same adsorption capacity. The mechanism of adsorption on $[Cu(L)_{1/3}(H_2O)]\cdot 8DMA$ has not been determined (Li et al. 2012): "The interactions between electron-rich organosulfur compounds and unsaturated metal site (coordination), p–p interactions between aromatic linker and conjugated organosulfur compounds, and so on."

The microporous titanium-containing MIL-125(Ti) is attractive for the adsorption and catalytic oxidation reactions in the liquid phase; however, its micropores are too small to allow the adsorption of the large molecules of aromatic sulfur compounds, such as DBT. Recently, the hierarchically structured micro-/mesoporous analog of MIL-125(Ti) was synthesized by vapor-assisted crystallization (McNamara and Hicks 2015) (Figure 10.5). This new MOF has been found to be more active than "conventional" microporous MIL-125(Ti) in the catalytic oxidation of DBT with tert-butyl hydroperoxide in solution, presumably due to the access the DBT molecule has to the active sites within the MOF. The mechanism of adsorption was studied using thiophene, the smallest molecule among aromatic sulfur compounds. Thiophene adsorption studies suggested that hierarchically structured MIL-125(Ti) has a higher number of the same kind of adsorption sites than its microporous counterpart (McNamara and Hicks 2015).

One can make the following conclusions. In the majority of reported studies, the mesoporous MOFs used for adsorption of the aromatic sulfur compound in solution contained transition metal (TM) cations as the CUS. There are only few papers that report adsorptive desulfurization on the mesoporous MOF with an Al(III) site (Maes et al. 2011, Van de Voorde et al. 2013), and the mechanism of the interaction of aromatic sulfur compounds with Al(III) is not known. So, the *first conclusion* is that while the presence of the TM CUS is assumed to be needed for adsorptive desulfurization, the mechanism of the interaction of aromatic sulfur compounds with the CUS in the mesoporous MOFs is still not known, and it needs to be investigated. The *second conclusion* is that a "satisfactory" adsorbed amount of the aromatic sulfur compound on the mesoporous MOFs was realized on the Cu(I) sites grafted onto the surface of MIL-100(Fe), with adsorption conducted from n-alkane as the solvent (Khan and Jhung 2012). Here, the adsorption site was believed to be the supported Cu(I), and the MOF served as an inert, high-surface-area mesoporous support material, rather than the sorbent per se, with its distinct adsorption sites. The *third conclusion* is that significant adsorbed amounts of aromatic sulfur compounds in the presence of aromatic hydrocarbons have not yet been achieved.

Furthermore, in the majority of reported studies, an alkane solvent in the model fuels was n-octane or n-heptane. The boiling point of n-octane is 125°C, while the boiling point of n-heptane is 98°C. Such n-alkane solvents with rather low boiling points were used to prepare model *gasoline*, which has a rather low boiling point range. However, commercially available gasoline actually contains mostly isoalkanes (such as isooctane), which are responsible for its high ON. Therefore, the reported short-chain n-alkanes used as solvents for model gasoline are *not* the representative solvents to model commercial gasoline to be desulfurized by adsorption. The *fourth conclusion* is that there is only one paper (Li et al. 2012) that reports adsorptive desulfurization on mesoporous MOF with model gasoline containing *iso*octane, that is, the kind of solvent representative of "real" gasoline.

On the other hand, model *diesel* and refinery oils in adsorption desulfurization studies are expected to have a long straight-chain alkane with a high cetane number and a high boiling point as the solvent. However, only one paper (Achmann et al. 2010) reports adsorption of the aromatic sulfur compound (thiophene) from n-alkane with a long straight-chain (dodecane) as the solvent; however, this study was conducted with *microporous* MOF. The *fifth conclusion* is that adsorptive desulfurization studies on *mesoporous* MOFs with model diesel or heating oil with a long-chain n-alkane as the solvent have not been reported.

We further note that fused-ring aromatic sulfur compounds have rather high boiling points, for example, BT has its boiling point at 221°C. Fused-ring aromatic sulfur compounds are concentrated in kerosene and oil refinery fractions, where long straight-chain alkanes are also concentrated such as, for instance, n-tetradecane with a boiling point at 253°C. Therefore, the long-chain hydrocarbon n-tetradecane, rather than n-octane or n-heptane, would be a suitable solvent for model diesel fuel, refinery, and heating oils. Then, the *sixth conclusion* is that an adsorption study on mesoporous

MOFs from model fuels, where both the target aromatic sulfur compound and the alkane solvent have high boiling temperatures (the typical oil), has not been reported.

The formation of adsorption complexes (McKinley and Angelici 2003, Zhu et al. 2013, Dai et al. 2014, Demir et al. 2014), in the system "liquid/solid" is usually assumed, but their direct spectroscopic characterization is very rare, as was noted by us (Samokhvalov and Tatarchuk 2010). The *seventh conclusion* is that the spectroscopic characterization of the adsorption complexes formed by aromatic sulfur compounds with mesoporous MOFs by molecular spectroscopy was not reported. This direction seems particularly interesting in the mechanistic studies of adsorptive sulfurization of liquid fuels with mesoporous MOFs.

REFERENCES

Achmann, S., G. Hagen, M. Hämmerle, I. M. Malkowsky, C. Kiener, and R. Moos. 2010. Sulfur removal from low-sulfur gasoline and diesel fuel by metal-organic frameworks. *Chemical Engineering & Technology* 33(2):275–280.

Ahmed, I., Z. Hasan, N. A. Khan, and S. H. Jhung. 2013a. Adsorptive denitrogenation of model fuels with porous metal-organic frameworks (MOFs): Effect of acidity and basicity of MOFs. *Applied Catalysis B: Environmental* 129:123–129.

Ahmed, I. and S. H. Jhung. 2014. Adsorptive denitrogenation of model fuel with CuCl-loaded metal-organic frameworks (MOFs). *Chemical Engineering Journal* 251:35–42.

Ahmed, I., N. A. Khan, Z. Hasan, and S. H. Jhung. 2013b. Adsorptive denitrogenation of model fuels with porous metal-organic framework (MOF) MIL-101 impregnated with phosphotungstic acid: Effect of acid site inclusion. *Journal of Hazardous Materials* 250:37–44.

Ahmed, I., N. A. Khan, and S. H. Jhung. 2013c. Graphite oxide/metal-organic framework (MIL-101): Remarkable performance in the adsorptive denitrogenation of model fuels. *Inorganic Chemistry* 52(24):14155–14161.

Babich, I. V. and J. A. Moulijn. 2003. Science and technology of novel processes for deep desulfurization of oil refinery streams: A review. *Fuel* 82(6):607–631.

Bhadra, B. N., K. H. Cho, N. A. Khan, D. Y. Hong, and S. H. Jhung. 2015. Liquid-phase adsorption of aromatics over a metal-organic framework and activated carbon: Effects of hydrophobicity/hydrophilicity of adsorbents and solvent polarity. *Journal of Physical Chemistry C* 119(47):26620–26627.

Boniek, D., D. Figueiredo, A. F. B. Santos, and M. A. Resende Stoianoff. 2015. Biodesulfurization: A mini review about the immediate search for the future technology. *Clean Technologies and Environmental Policy* 17(1):29–37.

Campos-Martin, J. M., M. C. Capel-Sanchez, P. Perez-Presas, and J. L. G. Fierro. 2010. Oxidative processes of desulfurization of liquid fuels. *Journal of Chemical Technology and Biotechnology* 85(7):879–890.

Cychosz, K. A., A. G. Wong-Foy, and A. J. Matzger. 2008. Liquid phase adsorption by microporous coordination polymers: Removal of organosulfur compounds. *Journal of the American Chemical Society* 130(22):6938–6939.

Dai, T., M. L. McKee, and A. Samokhvalov. 2014. Adsorption of naphthalene and indole on F300 MOF in liquid phase by the complementary spectroscopic, kinetic and DFT studies. *Journal of Porous Materials* 21(5):709–727.

Demir, M., M. L. McKee, and A. Samokhvalov. 2014. Interactions of thiophenes with C300 Basolite MOF in solution by the temperature-programmed adsorption and desorption, spectroscopy and simulations. *Adsorption* 20(7):829–842.

Dhage, P., A. Samokhvalov, D. Repala, E. C. Duin, and B. J. Tatarchuk. 2011. Regenerable Fe-Mn-ZnO/SiO₂ sorbents for room temperature removal of H₂S from fuel reformates: Performance, active sites, Operando studies. *Physical Chemistry Chemical Physics* 13(6):2179–2187.

Hernandez-Maldonaldo, A. J. and R. T. Yang. 2004. Desulfurization of transportation fuels by adsorption. *Catalysis Reviews: Science and Engineering* 46(2):111–150.

Huang, M. H., G. G. Chang, Y. Su, H. B. Xing, Z. G. Zhang, Y. W. Yang, Q. L. Ren, Z. B. Bao, and B. L. Chen. 2015. A metal-organic framework with immobilized Ag(I) for highly efficient desulfurization of liquid fuels. *Chemical Communications* 51(61):12205–12207.

Hwang, Y. K., D.-Y. Hong, J.-S. Chang, S. H. Jhung, Y.-K. Seo, J. Kim, A. Vimont, M. Daturi, C. Serre, and G. Férey. 2008. Amine grafting on coordinatively unsaturated metal centers of MOFs: Consequences for catalysis and metal encapsulation. *Angewandte Chemie International Edition* 47(22):4144–4148.

Ismagilov, Z., S. Yashnik, M. Kerzhentsev, V. Parmon, A. Bourane, F. M. Al-Shahrani, A. A. Hajji, and O. R. Koseoglu. 2011. Oxidative desulfurization of hydrocarbon fuels. *Catalysis Reviews—Science and Engineering* 53(3):199–255.

Ito, E. and J. van Veen. 2006. On novel processes for removing sulphur from refinery streams. *Catalysis Today* 116(4):446–460.

Jia, S. Y., Y. F. Zhang, Y. Liu, F. X. Qin, H. T. Ren, and S. H. Wu. 2013. Adsorptive removal of dibenzothiophene from model fuels over one-pot synthesized PTA@MIL-101(Cr) hybrid material. *Journal of Hazardous Materials* 262:589–597.

Khan, N. A., Z. Hasan, and S. H. Jhung. 2014. Ionic liquids supported on metal-organic frameworks: Remarkable adsorbents for adsorptive desulfurization. *Chemistry—A European Journal* 20(2):376–380.

Khan, N. A. and S. H. Jhung. 2012. Low-temperature loading of Cu⁺ species over porous metal-organic frameworks (MOFs) and adsorptive desulfurization with Cu⁺-loaded MOFs. *Journal of Hazardous Materials* 237–238:180–185.

Li, S. L., Y. Q. Lan, H. Sakurai, and Q. Xu. 2012. Unusual regenerable porous metal-organic framework based on a new triple helical molecular necklace for separating organosulfur compounds. *Chemistry—A European Journal* 18(51):16302–16309.

Li, Y. X., W. J. Jiang, P. Tan, X. Q. Liu, D. Y. Zhang, and L. B. Sun. 2015. What matters to the adsorptive desulfurization performance of metal-organic frameworks? *Journal of Physical Chemistry C* 119(38):21969–21977.

Maes, M., M. Trekels, M. Boulhout, S. Schouteden, F. Vermoortele, L. Alaerts, D. Heurtaux et al. 2011. Selective removal of N-heterocyclic aromatic contaminants from fuels by Lewis acidic metal-organic frameworks. *Angewandte Chemie International Edition* 50(18):4210–4214.

Martinez-Palou, R. and R. Luque. 2014. Applications of ionic liquids in the removal of contaminants from refinery feedstocks: An industrial perspective. *Energy & Environmental Science* 7(8):2414–2447.

McKinley, S. G. and R. J. Angelici. 2003. Deep desulfurization by selective adsorption of dibenzothiophenes on Ag⁺/SBA-15 and Ag⁺/SiO₂. *Chemical Communications* 20:2620–2621.

McNamara, N. D. and J. C. Hicks. 2015. Chelating agent-free, vapor-assisted crystallization method to synthesize hierarchical microporous/mesoporous MIL-125 (Ti). *ACS Applied Materials & Interfaces* 7(9):5338–5346.

Nuzhdin, A. L., K. A. Kovalenko, D. N. Dybtsev, and G. A. Bukhtiyarova. 2010. Removal of nitrogen compounds from liquid hydrocarbon streams by selective sorption on metal-organic framework MIL-101. *Mendeleev Communications* 20(1):57–58.

Pilloni, M., F. Padella, G. Ennas, S. R. Lai, M. Bellusci, E. Rombi, F. Sini et al. 2015. Liquid-assisted mechanochemical synthesis of an iron carboxylate Metal Organic Framework and its evaluation in diesel fuel desulfurization. *Microporous and Mesoporous Materials* 213:14–21.

Ryzhikov, A., I. Bezverkhyy, and J.-P. Bellat. 2008. Reactive adsorption of thiophene on Ni/ZnO: Role of hydrogen pretreatment and nature of the rate determining step. *Applied Catalysis B: Environmental* 84:766–772.

Samokhvalov, A. 2011. Heterogeneous photocatalytic reactions of sulfur aromatic compounds. *ChemPhysChem* 12(16):2870–2885.

Samokhvalov, A. 2012. Desulfurization of real and model liquid fuels using light: Photocatalysis and photochemistry. *Catalysis Reviews: Science and Engineering* 54(3):281–343.

Samokhvalov, A., E. C. Duin, S. Nair, M. Bowman, Z. Davis, and B. J. Tatarchuk. 2010a. Study of the surface chemical reactions of thiophene with Ag/titania by the complementary temperature-programmed electron spin resonance, temperature-programmed desorption, and x-ray photoelectron spectroscopy: Adsorption, desorption, and sorbent regeneration mechanisms. *Journal of Physical Chemistry C* 114(9):4075–4085.

Samokhvalov, A., E. C. Duin, S. Nair, and B. J. Tatarchuk. 2011. Adsorption and desorption of dibenzothiophene on Ag-titania studied by the complementary temperature-programmed XPS and ESR. *Applied Surface Science* 257(8):3226–3232.

Samokhvalov, A., S. Nair, E. C. Duin, and B. J. Tatarchuk. 2010b. Surface characterization of Ag/titania adsorbents. *Applied Surface Science* 256:3647–3652.

Samokhvalov, A. and B. J. Tatarchuk. 2010. Review of experimental characterization of active sites and determination of molecular mechanisms of adsorption, desorption and regeneration of the deep and ultradeep desulfurization sorbents for liquid fuels. *Catalysis Reviews: Science and Engineering* 52:381–410.

Senkovska, I., E. Barea, J. A. R. Navarro, and S. Kaskel. 2012. Adsorptive capturing and storing greenhouse gases such as sulfur hexafluoride and carbon tetrafluoride using metal-organic frameworks. *Microporous and Mesoporous Materials* 156:115–120.

Speight, J. G. 2000. *The Desulfurization of Heavy Oils and Residua*. Marcel Dekker Inc. New York.

Stanislaus, A., A. Marafi, and M. S. Rana. 2010. Recent advances in the science and technology of ultra low sulfur diesel (ULSD) production. *Catalysis Today* 153(1–2):1–68.

Tatarchuk, B., H. Yang, and S. Nair. 2008. Silver-based sorbents. Patent US 2008/0283446 A1.

Van de Voorde, B., M. Boulhout, F. Vermoortele, P. Horcajada, D. Cunha, J. S. Lee, J. S. Chang et al. 2013. N/S-heterocyclic contaminant removal from fuels by the mesoporous metal-organic framework MIL-100: The role of the metal ion. *Journal of the American Chemical Society* 135(26):9849–9856.

Wu, L. M., J. Xiao, Y. Wu, S. K. Xian, G. Miao, H. H. Wang, and Z. Li. 2014a. A combined experimental/computational study on the adsorption of organosulfur compounds over metal-organic frameworks from fuels. *Langmuir* 30(4):1080–1088.

Wu, Y., J. Xiao, L. M. Wu, M. Chen, H. X. Xi, Z. Li, and H. H. Wang. 2014b. Adsorptive denitrogenation of fuel over metal organic frameworks: Effect of N-types and adsorption mechanisms. *Journal of Physical Chemistry C* 118(39):22533–22543.

Yang, H., R. Sothen, D. R. Cahela, and B. J. Tatarchuk. 2008. Breakthrough characteristics of reformate desulfurization using ZnO sorbents for logistic fuel cell power systems. *Industrial & Engineering Chemistry Research* 47:10064–10070.

Zhang, H. X., H. L. Huang, C. X. Li, H. Meng, Y. Z. Lu, C. L. Zhong, D. H. Liu, and Q. Y. Yang. 2012. Adsorption behavior of metal-organic frameworks for thiophenic sulfur from diesel oil. *Industrial & Engineering Chemistry Research* 51(38):12449–12455.

Zhu, M., P. Northrup, C. Shi, S. J. L. Billinge, D. L. Sparks, and G. A. Waychunas. 2013. Structure of sulfate adsorption complexes on ferrihydrite. *Environmental Science & Technology Letters* 1(1):97–101.

11 Adsorption of Miscellaneous Organic Compounds from Nonaqueous Solutions

11.1 ADSORPTION OF ORGANIC COMPOUNDS ON MESOPOROUS MOFs

MIL-101(Cr) is one of the most well-investigated metal–organic frameworks (MOFs), and its adsorption in the nonaqueous phase has been studied extensively. Adsorption of polar ethylcinnamate (artificial flavor) versus nonpolar styrene (feedstock for industrial polymer synthesis) (Figure 11.1) was studied on activated MIL-101(Cr) (Henschel et al. 2011).

In addition, MIL-101(Cr) with supported palladium metal nanoparticles was prepared by impregnating the MOF with a chloroform solution of palladium acetylacetonate precursor, followed by reduction in hydrogen at 473 K to synthesize the MIL-101(Cr)–Pd with 1 wt.% Pd, which was expected to have the capability of catalytic hydrogenation (Henschel et al. 2011). This synthesis has been successful due to the known rather high chemical and thermal stability of MIL-101(Cr) (e.g., Liu et al. 2013). Adsorption of styrene or ethylcinnamate from solution on MIL-101(Cr)–Pd under argon at 298 K was studied as a step in catalytic hydrogenation (Henschel et al. 2011). The highest adsorbed amount of ethylcinnamate from n-heptane on MIL-101(Cr) was ca. 800 mg/g, and that on MIL-101(Cr)–Pd was about 600 mg/g. So, the deposition of Pd did not strongly decrease the adsorption capacity of the MOF. Furthermore, the adsorption capacity of the MOF under study was found to strongly depend on the polarity of the adsorbate, as well as the polarity of the adsorbate versus that of the solvent. Specifically, the adsorbed amount of polar ethylcinnamate on MIL-101(Cr)–Pd was 600 mg/g sorbent, while that of nonpolar styrene was only 50 mg/g. When the solvent was changed from the one with a low polarity (dielectric constant $\varepsilon = 1.9$–2.4) to the highly polar ethylacetate ($\varepsilon = 6.0$) or dichloromethane ($\varepsilon = 8.9$), the adsorbed amount of styrene decreased to nearly zero (Henschel et al. 2011).

Table 11.1 shows published data on the adsorption of miscellaneous aromatic and heterocyclic compounds on MIL-101 in nonwater solvents.

Fullerenes represent one of the most intriguing classes of inorganic carbon molecules (Taylor and Walton 1993), and they have been intensively studied due to their unusual properties. Unfortunately, practical applications of fullerenes are limited, in part because the separation of C60 from C70 fullerenes is difficult. The adsorption

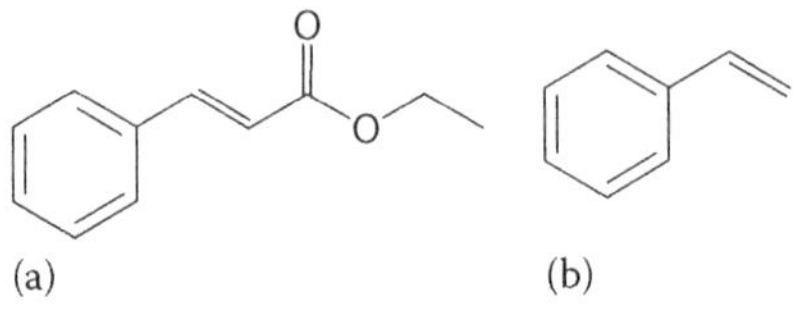

(a)   (b)

FIGURE 11.1   Structures of (a) ethylcinnamate and (b) styrene.

of C60 and C70 fullerenes from toluene solution at 30°C on MIL-101(Cr) activated at 150°C for 12 h followed a pseudo-second-order kinetic rate law (Yang and Yan 2012). The adsorption kinetic rate constant for the C70 fullerene was three to five times higher than that for the C60 fullerene. The highest adsorption capacity for C70 from the toluene solution at 30°C was 198.4 mg/g sorbent, which was about 29-fold higher than that of C60. The adsorption of fullerenes follows the Langmuir model; the adsorption of fullerenes on MIL-101(Cr) was found to be controlled by the change of entropy, ΔS (Yang and Yan 2012). The thermodynamic model used by the authors assumes that adsorption can be viewed as occurring in the two sequential steps: (1) desolvation of the fullerene molecule with the release of solvent molecules at $\Delta H_1 > 0$ and $\Delta S_1 > 0$, and (2) adsorption of the desolvated fullerene on MIL-101(Cr) at $\Delta H_2 < 0$ and $\Delta S_2 < 0$. This model explains the preferential adsorption of the larger C70 versus the smaller C60. The "spent" MIL-101(Cr) sorbent was regenerated by washing with o-dichlorobenzene, p-xylene, or toluene under ultra-sonication; o-dichlorobenzene yielded the quickest desorption of C70 (Yang and Yan 2012). Therefore, the more polar solvent causes faster desorption of adsorbed fullerenes, which are nonpolar compounds. These findings are consistent with the decrease in the adsorbed amount of nonpolar styrene on MIL-101(Cr), as mentioned earlier, when the polar solvent was used (Henschel et al. 2011). MIL-101(Cr) was also used to adsorb fullerenes larger than C70 from the extract of carbon soot in toluene (Yang and Yan 2012). The larger fullerenes were found to be able to replace the C70 preadsorbed on MIL-101(Cr); therefore, the larger fullerenes have higher affinity toward MIL-101(Cr) than toward C70, and C70 has higher affinity than C60. The mechanism of adsorption of fullerenes was not experimentally determined, but one can speculate that π–π interactions may be significant, which explains the better adsorption of the larger fullerenes.

Chlorinated aromatic compounds are widely used as herbicides; many of them have an offensive odor and are toxic or even carcinogenic (Alavanja et al. 2004). Chloroaromatic pesticides can easily accumulate in soil and water, which represents a global environmental concern (e.g., Pereira et al. 1996, Muller et al. 2008). Among the chlorinated aromatic compounds, polychlorinated dibenzodioxins (PCDDs) are most well known for their acute toxicity. The defoliant Agent Orange (Stellman et al. 2003) sprayed over vegetation during the Vietnam War contained PCDDs, which caused multiple birth defects and cancer. Brominated aromatic compounds produced by burning chemical waste (Schafer and Ballschmiter 1986) are also very toxic (Behnisch et al. 2003). Therefore, it is essential to find active and selective sorbents that can effectively remove chlorinated and brominated aromatic compounds.

TABLE 11.1

Adsorption and Desorption of Miscellaneous Aromatic Compounds from Nonaqueous Solvents on MIL-101

MOF	Adsorbate	Use	Formula	Solvent	Adsorbed Amount and Bonding	Reference
MIL-101(Cr)	Ethylcinnamate	Flavoring	Figure 11.1	n-Heptane	800 mg/g; polar	Henschel et al. (2011)
MIL-101(Cr)-Pd	Ethylcinnamate	Flavoring	Figure 11.1	n-Heptane	600 mg/g; polar	Henschel et al. (2011)
MIL-101(Cr)	Styrene	Synthesis of polymers	Figure 11.1	n-Heptane	50 mg/g; polar	Henschel et al. (2011)
MIL-101(Cr)	C70 fullerene	—		Toluene	198.4 mg/g; π–π in adsorption. Polarity in desorption	Yang and Yan (2012)
MIL-101(Cr)	C60 fullerenes	—		Toluene	6.76 mg/g; π–π and polarity	Yang and Yan (2012)

Adsorption of several chlorinated aromatic compounds that model the PCDDs (Figure 11.2) was investigated on activated MIL-101(Cr) in 1,3,5-triisopropylbenzene (TIPB) as the solvent at 293, 298, and 303 K (Kim et al. 2013).

Prior to adsorption, MIL-101(Cr) was activated at 373 K under vacuum overnight. The adsorption kinetics followed the pseudo-second-order rate law, and the adsorption equilibrium fits the Langmuir model (Kim et al. 2013). The adsorption of 2-chloroanisole on MIL-101(Cr) was found to be spontaneous and exothermic. The adsorption rate constant increases with the temperature. The mechanism of adsorption was tentatively defined as the π–π interaction between the chlorinated aromatic compounds and the benzene rings in the linker of the MOF, although no mechanistic study was conducted. The role of the coordinatively unsaturated site (CUS) of the activated MIL-101(Cr) in adsorption was not determined. To determine competitive adsorption, MIL-101(Cr) was left in contact with an equimolar solution containing the mixture of chlorobenzene, 2-chlorotoluene, 1,3-dichlorobenzene, and 2-chloroanisole at 303 K. The 2-chloroanisole showed preferential adsorption among other chlorinated aromatic compounds within the concentration range of 0.2–1.0 M. This effect was explained by the larger size of the molecule and its presumed stronger interaction with the interior of the mesopores of the MOF, tentatively by the polar OCH_3 groups (Kim et al. 2013).

An active pharmaceutical ingredient (API) is a chemical substance used in manufacturing medicinal drugs. Adsorption of several APIs (Figure 11.3) was studied on commercial mesoporous MOF Basolite F300 (Centrone et al. 2011), which is similar to MIL-100(Fe). The molecules of the reported APIs have the phenyl group and the following substituents: keto (I), hydroxyl (II), aliphatic secondary amine (III), and the ternary amine (IV) group.

Before adsorption, Basolite F300 MOF was activated in vacuum at 170°C for 12 h; the adsorbed water and an excess of 1,3,5-benzenetricarboxylic (BTC) remaining from the synthesis of the MOF were removed, and Fe(III) CUSs were generated. When

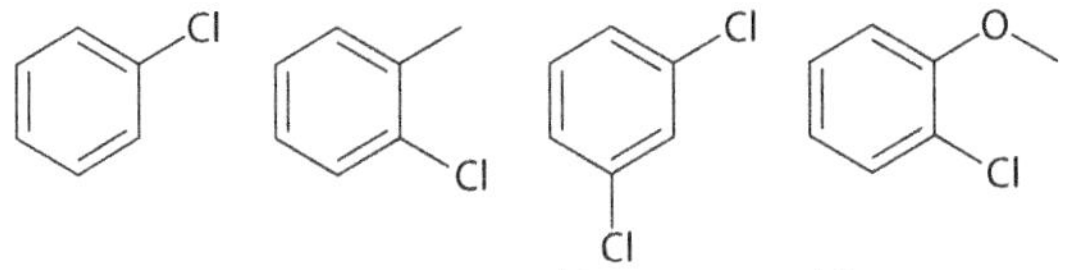

(a)   (b)   (c)   (d)

FIGURE 11.2   Chlorinated aromatic compounds as in Kim et al. (2013): (a) chlorobenzene, (b) 2-chlorotoluene, (c) 1,3-dichlorobenzene, and (d) 2-chloroanisole.

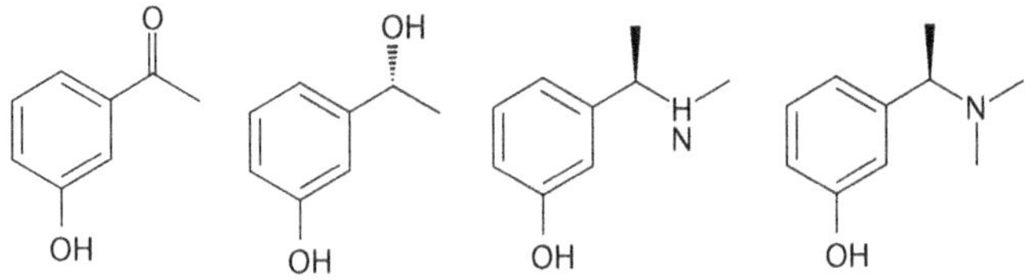

I   II   III   IV

FIGURE 11.3   The APIs used in the study. (From Centrone, A. et al., Small, 7(16), 2356, 2011.)

activated Basolite F300 was brought in contact with the methanol solution of the APIs III and IV (Figure 11.3), which contain amino groups, Basolite F300 changed its color, which indicates the formation of coordination compounds (Centrone et al. 2011). In contrast, there was no change in the color of Basolite F300 MOF when in contact with the solution of APIs with ketone and alcohol groups (compounds I and II in Figure 11.3). After adsorption of amine APIs on Basolite F300, the MOF was filtered and its UV-Vis diffuse reflectance spectrum, DRS was measured. An increase in absorption at 250–700 nm suggested the formation of coordination bonds between the Fe(III) in the MOF (electron acceptor) and amino groups in compounds III and IV (electron donor). After exposure to methanol, Basolite F300 MOF showed the Fourier transform infrared (FTIR) peak at 1710 cm^{-1}, which was assigned to the OH bending mode of methanol molecules forming a complex with Fe(III) CUS (Centrone et al. 2011).

Fused-ring aromatic and heterocyclic compounds are abundantly present in virtually all fossils and fossil fuels. The structure of the prototype fused-ring aromatic hydrocarbon naphthalene is similar to that of indole. As was reported in our recent publication (Dai et al. 2014), naphthalene forms a 1:1 stoichiometric adsorption complex with activated commercial mesoporous MOF Basolite F300 when prepared in the "matrix" of solid eicosane. Herein, the eicosane matrix models the long-chain alkanes abundantly present in solid fossils such as "heavy" petroleum, Canadian bitumen, and the U.S. shale and tar sands. In this adsorption complex, the molecule of naphthalene is confined within the mesocavity of Basolite F300 MOF as has been shown by vibronically resolved fluorescence spectroscopy. The adsorbed naphthalene molecule is very weakly electronically bound to Fe(III) CUS as determined by the density functional theory (DFT) calculations and is strongly dispersively stabilized by side interactions with the benzene rings of the BTC linker of Basolite F300 (Dai et al. 2014).

An interesting new potential application area for mesoporous MOFs based on adsorption of aromatic hydrocarbons is adsorptive dearomatization of refinery streams to diesel fuel, where an increase in the cetane number is needed. While the adsorptive dearomatization of refinery streams was studied with ionic liquids (e.g., Martinez-Palou and Luque 2014), an adsorptive dearomatization of petroleum refinery streams by mesoporous MOFs remains unexplored.

Biomass is thought to partially replace fossils in the production of sustainable fine chemicals, polymers, and fuels. The major classes of organic compounds present in "biooil" obtained at biorefineries from plant biomass have been determined (Fernando et al. 2006), and promising "platform molecules" have been identified (Tong et al. 2010). The platform furans from plant biomass have one or two substituents in the five-member heteroaromatic ring with the oxygen atom. Furfural from plant biomass (Salak Asghari and Yoshida 2006) has the potential of becoming a platform chemical (Cai et al. 2014) with applications in the production of sustainable resins (Yoshida et al. 2008), lubricants, adhesives, flavoring agents (Adams et al. 1997), etc. Furoic acid can be used in the synthesis of sustainable polymers (Reeb and Chauvel 1987), and furfuryl alcohol is used as an industrial solvent (Chheda et al. 2007) and feedstock for synthesis of poly-furanol (Liu et al. 2005). The compounds 2,5-furan-dicarboxylic acid (2,5-FDCA) and 5-hydroxymethyl furfural (5-HMF) were termed "sleeping giants" of renewable chemicals and fuels obtained from plant biomass (Tong et al. 2010); 2,5-FDCA is the sustainable equivalent of terephthalic acid that can be used in the synthesis of polymers (Munoz et al. 2011). To our knowledge and on the basis of intensive research to produce sustainable fine chemicals, plastics, and fuels (Roman-Leshkov et al. 2007) based on furans (Chheda et al. 2007), the purification and separation of sustainable fuels and chemicals using mesoporous MOFs have not been reported. We believe that this area is ripe for exploration.

11.2 ADSORPTION OF LARGE-MOLECULE BIOLOGICALLY ACTIVE ORGANIC COMPOUNDS ON MESOPOROUS MOFs

Molecules of many biologically active compounds such as DNA, proteins, and enzymes are much larger than the typical size of microporous channels which lead to mesocavities in mesoporous MOFs. For adsorption of biologically active organic compounds, MOFs with extralarge pore sizes are needed. One can highlight the two specific cases of adsorption that are unique to mesoporous MOFs. First, if the large molecule cannot diffuse to the adsorption site (metal or the linker) located in the interior of the mesopore or at the microporous "window," the adsorbate molecule interacts only with the outer surface of the MOF crystals, as was indeed reported for adsorption of DNA (Fang et al. 2014, Guo et al. 2014, Wang et al. 2014). The second case is when the adsorbate molecule can diffuse through the microporous "windows" of the MOF toward the mesopores, and it can therefore interact with metal and/or linker units located in the mesopores or the "windows." In the latter case, the peculiarities of the structure of the given MOF play a major role in determining its adsorption capacity and selectivity. Some of the most typical adsorption material systems of this second type are discussed here (Table 11.2).

TABLE 11.2
Adsorption of Biologically Active Compounds on Mesoporous MOFs

MOF	Adsorbate	Molecular Structure of the Adsorbate	Encapsulation	Reference
MIL-101(Cr)	Indomethacin	Figure 11.4	THF	Čendak et al. (2014)
MIL-101(Fe)	Indomethacin	Figure 11.4	THF	Čendak et al. (2014)
MIL-101(Fe)-NH$_2$	Indomethacin	Figure 11.4	THF	Čendak et al. (2014)
MIL-101(Al)-NH$_2$	Indomethacin	Figure 11.4	THF	Čendak et al. (2014)
IRMOF-74-IV	Vitamin B12	Figure 11.5	Methanol, [vitamin B12]$_0$ = 1×10^{-4} M	Deng et al. (2012)
IRMOF-74-V	MOP-18		Chloroform, [MOP-18]$_0$ = 1×10^{-5} M	Deng et al. (2012)
3, see Li et al. (2015)	Doxorubicin		DMF	Li et al. (2015)

Indomethacin (Figure 11.4) is a nonsteroidal antiinflammatory drug (NSAID). The adsorption of indomethacin on several MIL-101 MOFs was studied (Čendak et al. 2014) from the solution in tetrahydrofuran (THF).

The interactions of the molecules of adsorbed indomethacin and the molecules of coadsorbed THF solvent with the metal sites and the linkers in the MOFs were investigated by solid-state nuclear magnetic resonance (NMR) spectroscopy (Čendak et al. 2014). Prior to the adsorption, the sorbents MIL-101 were "activated" using organic solvents. Specifically, MIL-101(Al)–NH_2 was activated overnight in methanol at room temperature. MIL-101(Fe) was activated in N,N-dimethylformamide (DMF) at 333 K, followed by drying at 333 K and treatment in acetone at 323 K. The activation of MIL-101(Fe)–NH_2 was conducted in DMF at room temperature. Thus, standard thermal activation of MIL-101 that yield the CUS was not employed, and one cannot expect the formation of coordination compounds of the adsorbed indomethacin with the CUS of the metal. The adsorbed amounts were found to be quite high, between 0.9 and 1.1 g of indomethacin per 1 g of MIL-101. NMR measurements showed that for all metal cations in the MIL-101 in the study (Cr, Fe, or Al) and the two kinds of linkers (1,4-benzenedicarboxylic acid [BDC] or amino-BDC) in the respective MOF sorbents, the molecules of adsorbed indomethacin did not form chemical bonds with the sites of the MOFs. In addition, the intermolecular interactions between the molecules of adsorbed indomethacin that are located within the mesocavities were also not detected (Čendak et al. 2014). This study illustrates that mechanistic experimental research on identifying the type of chemical bonding between the MOF and the large-molecule adsorbates may become quite complicated. It seems likely that instead of one specific adsorption site with its distinct molecular mechanism, several adsorption sites are engaged simultaneously. The energy of each individual interaction is rather weak, and the identity of the adsorption interaction becomes elusive, even when rather sophisticated experimental methods are employed.

A new strategy to significantly increase the pore sizes in mesoporous MOFs up to >32 Å was reported (Deng et al. 2012). The isoreticular series of MOF-74 have been synthesized with linker units of variable molecular sizes. In these isoreticular MOFs (IRMOFS), the linker units contain 1 phenylene ring or 3, 4, 5, 6, 7, 9, 11 conjugated phenylene rings. The obtained IRMOFs have been termed IRMOF-74-I to XI, and they have variable pore apertures from 14 to 98 Å. Isoreticular IRMOF-74-I to IRMOF-74-XI have noninterpenetrating 3D structures, permanent porosity,

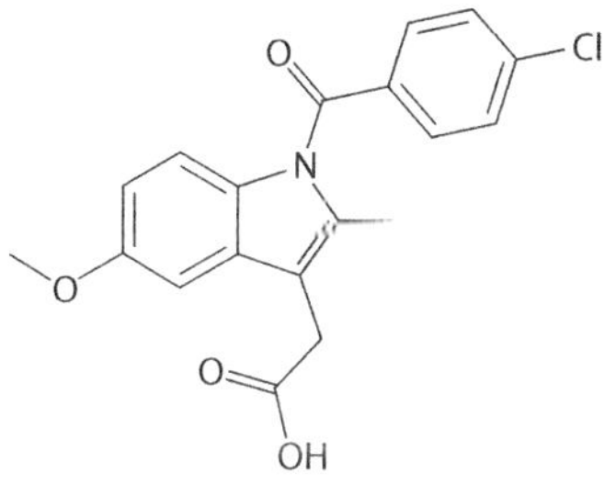

FIGURE 11.4 Molecular structure of indomethacin.

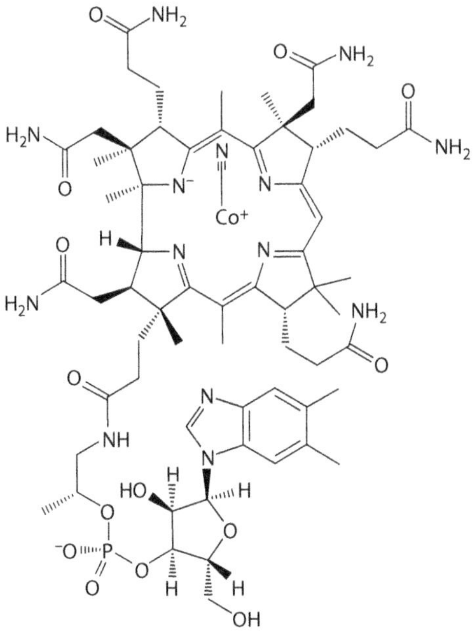

FIGURE 11.5 Molecular structure of vitamin B12.

and thermal stability up to 300°C. IRMOF-74-IV has a pore size of 34 Å (Deng et al. 2012). The molecule of vitamin B12 has the largest molecular dimension of 27Å (see Figure 11.5). The crystals of thermally activated IRMOF-74-IV were immersed in a 0.11 mM solution of vitamin B12 in methanol. Adsorption proceeded for up to 48 h as determined by UV-Vis spectrophotometry of vitamin B12 in solution through its characteristic absorbance at 480 nm. After inclusion of vitamin B12 into IRMOF-74-IV by adsorption in solution, the crystallinity of IRMOF-74-IV was maintained, as was found by comparison of the powder x-ray diffraction (XRD) patterns before and after adsorption. In contrast, IRMOF-74-III has a pore size slightly smaller than the size of vitamin B12 molecules. Indeed, adsorption of vitamin B12 on IRMOF-74-III did not proceed under the same adsorption conditions. Therefore, the mechanism of adsorption was explained by the spatial "match" between the molecular size of the adsorbate and the size of the mesopore in the MOF (Deng et al. 2012). In Table 11.2, MOP-18 denotes metal–organic rhombicuboctahedron cluster with a diameter of 34 Å which was studied in adsorption on IRMOF-74-V with a similar mesopore size.

The three new MOFs denoted as 1, 2, and 3 in Figure 11.6 were synthesized utilizing the reaction of the respective precursors of the ligands H_6L1, H_6L2, and H_6L3 with Zn(II) (Li et al. 2015).

The size of the inner sphere for both the T-Td and T-Oh polyhedrons increases from 13.0 and 20.0 Å in the MOF 1 to 14.0 and 23.0 Å in MOF 2, and to 15.0 and 25.0 Å in the respective MOF 3 (Figure 11.6). As proof of principle, the adsorption of cationic versus anionic dyes was studied on isoreticular mesoporous MOFs

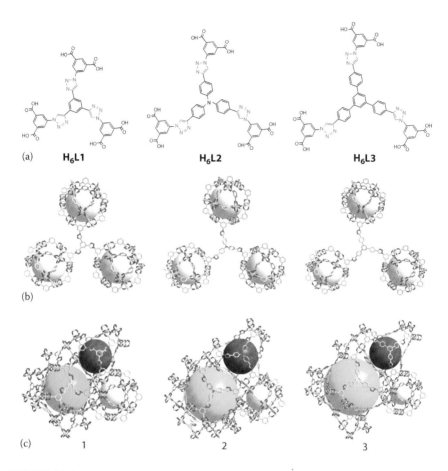

FIGURE 11.6 (a) Structure of the click-extended dendritic hexacarboxylate ligands **H₆L1**, **H₆L2**, and **H₆L3**. (b) (3,24)-Connected rht-type frameworks of MOFs 1, 2, and 3. (c) Three types of metal–organic polyhedrons in MOFs 1, 2, and 3: cub-Oh (yellow), T-Td (violet), and T-Oh (green). (From Li, P.Z., Wang, X.J., Tan, S.Y., Ang, C.Y., Chen, H.Z., Liu, J., Zou, R.Q., and Zhao, Y.L.: Clicked isoreticular metal-organic frameworks and their high performance in the selective capture and separation of large organic molecules. *Angew. Chemie Int. Ed.* 2015. 54(43). 12748–12752. Copyright Wiley-VCH Verlag GmbH & Co. KGaA. Reproduced with permission.)

1, 2, and 3 (Li et al. 2015). When the activated crystals of each MOF were immersed in the solutions of dyes methylene blue (MeB), rhodamine 6G (R6G), and brilliant blue R (BBR) in DMF, the color changes of the MOF powders were visible by the eye, apparently due to the penetration of the respective dye molecules within the MOF. With the relatively small MeB molecule, the changes in the color were observed with crystals of all three MOFs. On the other hand, with the larger R6G molecule, the color changed only in the crystals of MOFs 2 and 3 having the larger mesopore sizes. Adsorption of the largest dye molecule (BBR) occurred only with

MOF 3. Finally, after immersing the crystals of activated MOF 3 into a solution of doxorubicin (DOX) in DMF, the change of color of the MOF became visible. Therefore, MOF 3 has some potential in drug-delivery applications (Li et al. 2015).

REFERENCES

Adams, T. B., J. Doull, J. I. Goodman, I. C. Munro, P. Newberne, P. S. Portoghese, R. L. Smith et al. 1997. The FEMA GRAS assessment of furfural used as a flavour ingredient. *Food and Chemical Toxicology* 35(8):739–751.

Alavanja, M. C. R., J. A. Hoppin, and F. Kamel. 2004. Health effects of chronic pesticide exposure: Cancer and neurotoxicity. *Annual Review of Public Health* 25(1):155–197.

Behnisch, P. A., K. Hosoe, and S. Sakai. 2003. Brominated dioxin-like compounds: In vitro assessment in comparison to classical dioxin-like compounds and other polyaromatic compounds. *Environment International* 29(6):861–877.

Cai, C. M., T. Zhang, R. Kumar, and C. E. Wyman. 2014. Integrated furfural production as a renewable fuel and chemical platform from lignocellulosic biomass. *Journal of Chemical Technology & Biotechnology* 89(1):2–10.

Čendak, T., E. Žunkovič, T. U. Godec, M. Mazaj, N. Z. Logar, and G. Mali. 2014. Indomethacin embedded into MIL-101 frameworks: A solid-state NMR study. *Journal of Physical Chemistry C* 118(12):6140–6150.

Centrone, A., E. E. Santiso, and T. A. Hatton. 2011. Separation of chemical reaction intermediates by metal–organic frameworks. *Small* 7(16):2356–2364.

Chheda, J. N., G. W. Huber, and J. A. Dumesic. 2007. Liquid-phase catalytic processing of biomass-derived oxygenated hydrocarbons to fuels and chemicals. *Angewandte Chemie International Edition* 46(38):7164–7183.

Dai, J., M. L. McKee, and A. Samokhvalov. 2014. Adsorption of naphthalene and indole on F300 MOF in liquid phase by the complementary spectroscopic, kinetic and DFT studies. *Journal of Porous Materials* 21(5):709–727.

Deng, H. X., S. Grunder, K. E. Cordova, C. Valente, H. Furukawa, M. Hmadeh, F. Gandara et al. 2012. Large-pore apertures in a series of metal-organic frameworks. *Science* 336(6084):1018–1023.

Fang, J. M., F. Leng, X. J. Zhao, X. L. Hu, and Y. F. Li. 2014. Metal-organic framework MIL-101 as a low background signal platform for label-free DNA detection. *Analyst* 139(4):801–806.

Fernando, S., S. Adhikari, C. Chandrapal, and N. Murali. 2006. Biorefineries: Current status, challenges, and future direction. *Energy & Fuels* 20(4):1727–1737.

Guo, J. F., C. M. Li, X. L. Hu, C. Z. Huang, and Y. F. Li. 2014. Metal-organic framework MIL-101 enhanced fluorescence anisotropy for sensitive detection of DNA. *RSC Advances* 4(18):9379–9382.

Henschel, A., I. Senkovska, and S. Kaskel. 2011. Liquid-phase adsorption on metal-organic frameworks. *Adsorption* 17(1):219–226.

Kim, H. Y., S. N. Kim, J. Kim, and W. S. Ahn. 2013. Liquid phase adsorption of selected chloroaromatic compounds over metal organic frameworks. *Materials Research Bulletin* 48(11):4499–4505.

Li, P. Z., X. J. Wang, S. Y. Tan, C. Y. Ang, H. Z. Chen, J. Liu, R. Q. Zou, and Y. L. Zhao. 2015. Clicked isoreticular metal-organic frameworks and their high performance in the selective capture and separation of large organic molecules. *Angewandte Chemie International Edition* 54(43):12748–12752.

Liu, J., H. Wang, S. Cheng, and K.-Y. Chan. 2005. Nafion–polyfurfuryl alcohol nanocomposite membranes for direct methanol fuel cells. *Journal of Membrane Science* 246(1):95–101.

Liu, Q., L. Q. Ning, S. D. Zheng, M. N. Tao, Y. Shi, and Y. He. 2013. Adsorption of carbon dioxide by MIL-101(Cr): Regeneration conditions and influence of flue gas contaminants. *Scientific Reports* 3:Article number 2916.

Martinez-Palou, R. and R. Luque. 2014. Applications of ionic liquids in the removal of contaminants from refinery feedstocks: An industrial perspective. *Energy & Environmental Science* 7(8):2414–2447.

Muller, B., M. Berg, Z. P. Yao, X. F. Zhang, D. Wang, and A. Pfluger. 2008. How polluted is the Yangtze River? Water quality downstream from the Three Gorges Dam. *Science of the Total Environment* 402(2–3):232–247.

Munoz, D. E. D. C., M. A. Dam, and G. J. M. Gruter. 2011. Method for the preparation of 2,5-furandicarboxylic acid and for the preparation of the dialkyl ester of 2,5-furandicarboxylic acid. Patent WO2011043661 A1.

Pereira, W. E., J. L. Domagalski, F. D. Hostettler, L. R. Brown, and J. B. Rapp. 1996. Occurrence and accumulation of pesticides and organic contaminants in river sediment, water and clam tissues from the San Joaquin River and tributaries, California. *Environmental Toxicology and Chemistry* 15(2):172–180.

Reeb, R. and B. Chauvel. 1987. Crosslinkable polymers based on unsaturated esters of furoic acid and process for their preparation and application in the manufacture of coatings. Patent US4663411 A.

Roman-Leshkov, Y., C. J. Barrett, Z. Y. Liu, and J. A. Dumesic. 2007. Production of dimethylfuran for liquid fuels from biomass-derived carbohydrates. *Nature* 447(7147):982–985.

Salak Asghari, F. and H. Yoshida. 2006. Acid-catalyzed production of 5-hydroxymethyl furfural from d-fructose in subcritical water. *Industrial & Engineering Chemistry Research* 45(7):2163–2173.

Schafer, W. and K. Ballschmiter. 1986. Monobromo-polychloro-derivatives of benzene, biphenyl, dibenzofurane and dibenzodioxine formed in chemical-waste burning. *Chemosphere* 15(6):755–763.

Stellman, J. M., S. D. Stellman, R. Christian, T. Weber, and C. Tomasallo. 2003. The extent and patterns of usage of Agent Orange and other herbicides in Vietnam. *Nature* 422(6933):681–687.

Taylor, R. and D. R. M. Walton. 1993. The chemistry of fullerenes. *Nature* 363(6431):685–693.

Tong, X., Y. Ma, and Y. Li. 2010. Biomass into chemicals: Conversion of sugars to furan derivatives by catalytic processes. *Applied Catalysis A: General* 385(1–2):1–13.

Wang, G. Y., C. Song, D. M. Kong, W. J. Ruan, Z. Chang, and Y. Li. 2014. Two luminescent metal-organic frameworks for the sensing of nitroaromatic explosives and DNA strands. *Journal of Materials Chemistry A* 2(7):2213–2220.

Yang, C. X. and X. P. Yan. 2012. Selective adsorption and extraction of C70 and higher fullerenes on a reusable metal-organic framework MIL-101(Cr). *Journal of Materials Chemistry* 22(34):17833–17841.

Yoshida, N., N. Kasuya, N. Haga, and K. Fukuda. 2008. Brand-new biomass-based vinyl polymers from 5-hydroxymethylfurfural. *Polymer Journal* 40(12):1164–1169.

12 Encapsulation and Release of Medicinal Drugs by Mesoporous MOFs

12.1 THERAPY WITH ENGINEERED NANOPARTICLES

The therapy by engineered nanoparticles has been studied for longer than a decade (Samuel et al. 2003). There are many concerns about the cellular toxicity of nano-engineered materials. The first case is when one administers nondegradable engineered nanoparticles carrying an encapsulated (adsorbed) form of medicinal drug systemically (intravenously or per orally). After controlled release of the drug, the engineered nondegradable nanoparticles remain in the cellular medium and their long-term health effects are usually not known. Perhaps the most well-known case of toxicity of engineered nanoparticles is argyria, which was reported after ingestion of colloidal silver bound to a protein carrier (White et al. 2003). The ingested silver metal nanoparticles form complexes with cellular proteins and diffuse to the bloodstream, and further to the skin. Then, the silver–protein complexes are deposited within the skin and photoreduced by ambient sunlight to form highly dispersed metallic silver. The second case is when engineered nanoparticles carrying the encapsulated drug are biodegradable. In this case, one needs to ensure that the degradation products of the nanocarrier of the engineered drug are relatively safe: (1) the complex "nanoparticles–drug" needs to be destroyed *inside* the targeted cell or tissue, thus releasing the drug locally, and (2) after drug release, the engineered nanoparticle needs to be destroyed, preferably *outside* the targeted cell or tissue, so as not to compromise the desired local therapeutic effects. Therefore, the design of an efficient and safe form of the drug encapsulated on engineered nanoparticles needs a careful case-by-case study.

12.2 MESOPOROUS MOFs AS DRUG CARRIERS

Mesoporous metal–organic frameworks (MOFs) are suitable sorbents for adsorption/encapsulation and controlled desorption/release of pharmaceutical drugs, since MOFs can meet many of the essential requirements. First, adsorption and desorption of many biologically active organic compounds on mesoporous MOFs can be controlled by pH. This allows the controlled release of an encapsulated drug at certain intra- or intercellular values of the pH as found in the particular target tissues.

Second, certain mesoporous MOFs, especially MIL-100(Fe), MIL-101(Fe), MIL-100(Al), and MIL-101(Al), have been shown to be of relatively low toxicity since these MOFs contain both the nontoxic linker and the nontoxic metal. To ensure that the complex "MOF/encapsulated drug" can be delivered systemically, suitable nano-colloidal forms of mesoporous MOFs have been developed and termed nano-MOFs or NMOFs. The encapsulation of certain medicinal drugs on NMOFs has been investigated in the recent decade for at least three major classes of biologically active compounds: (1) anticancer drugs, (2) antiviral drugs, and (3) miscellaneous pharmaceuticals and personal care products (PPCPs).

12.3 ENCAPSULATION AND CONTROLLED DESORPTION OF ANTICANCER SMALL-MOLECULE DRUGS

In recent years, novel promising methods for treating cancer, for example, cancer immunotherapy using proteins, have been developed (Zhang and Sentman 2011). Still, the "mainstream" in cancer chemotherapy is using small-molecule heterocyclic drugs. Anticancer drugs can be considered life-saving. Therefore, some moderate toxicity of the drug carrier, the mesoporous MOF in this case, present in the encapsulated form together with the loaded anticancer drug is acceptable, given the life-saving therapeutic effects.

Table 12.1 shows publications on the encapsulation of small-molecule anticancer drugs on the mesoporous MOFs of the MIL-100 family and their controlled release.

Doxorubicin (DOX), Figure 12.1 belongs to the group of anthracycline antibiotics (Aubel-Sadron and Londos-Gagliardi 1984), although it is also widely used as an anticancer drug. Doxorubicin is believed to work by intercalation into the DNA of the cancer cells, thus preventing their division and tumor growth (Yokochi and Robertson 2004). Besides mesoporous MOFs, encapsulation of doxorubicin has been widely investigated on both inorganic nanoparticle-based (Li et al. 2012) and polymer (Zhan et al. 2013) carriers.

The molecule of doxorubicin has dimensions of about 15.3×11.9 Å. Encapsulation of DOX was studied on nanocrystalline MIL-100(Fe) or NMOF (Horcajada et al. 2010). Specifically, doxorubicin was encapsulated on the nano-MIL-100(Fe) from a solution of DOX hydrochloride in water for 24 h at room temperature. Then the nano-MIL-100(Fe) with the encapsulated drug was collected by centrifugation and dried in vacuum. MIL-100(Fe) has a small mesocage of 25 Å in diameter with pentagonal windows at 4.8×5.8 Å, and a large mesocage of 29 Å in diameter with both the larger hexagonal windows, at 8.6 Å, and the pentagonal windows (Horcajada et al. 2006). Thus, it was doubtful whether the DOX molecule would be able to enter the mesocavity of MIL-100 through any of the microporous windows. Indeed, the loading of DOX was found to be only 9.1 wt.% (by the mass of the MOF), which was explained by surface adsorption onto the nanocrystals of the NMOF, rather than by adsorption within the mesopores. The release of adsorbed doxorubicin into phosphate-buffered saline (PBS) at 37°C was achieved at about 100% yield in 14 days (Horcajada et al. 2010). Such long-term drug release is beneficial for proposed medicinal applications.

TABLE 12.1

Encapsulation and Controlled Desorption/Release of Anticancer Drugs from MIL-100

MOF	Drug Name	Drug Formula	Encapsulation	Host/Guest Bonding	Release	Reference
Nano-MIL-100(Fe)	Doxorubicin (DOX)	Figure 12.1	Activated MOF, water	On the surface of nano-MOF	PBS, 37°C, 100%, 14 days	Horcajada et al. (2010)
Nano-MIL-100(Fe)	Busulfan	Figure 12.2	Organic solvents	—	Cell culture	Horcajada et al. (2010)
Nano-MIL-100(Fe)	Topotecan	Figure 12.3	Nonactivated MOF, water	Coordination	Water at pH ≈ 6.7; PBS; photo-assisted release	di Nunzio et al. (2014)

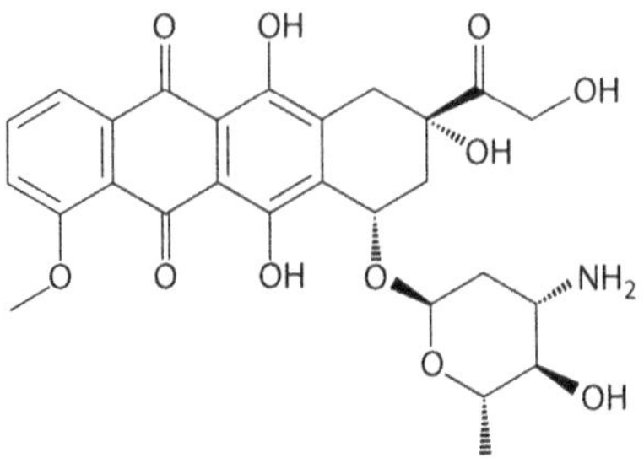

FIGURE 12.1   Molecular formula of doxorubicin (DOX) drug.

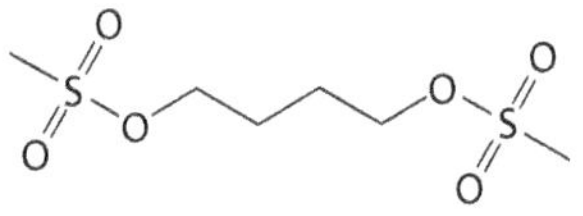

FIGURE 12.2   Molecular formula of busulfan drug.

The amphiphilic antitumoral drug busulfan (Bu) (Figure 12.2) is widely used in chemotherapy for leukemia because it represents an alternative to total-body irradiation.

Unfortunately, busulfan has rather poor stability in aqueous solutions and high hepatic toxicity (Goekkurt et al. 2007). Thus, it is important to design the treatment regime with busulfan based on a gradual release of the drug from a suitable carrier. Busulfan was adsorbed into the pores of NMOF MIL-100(Fe) from the suspension in nonwater solvents that included acetone, acetonitrile, chloroform, and dichloromethane at room temperature under stirring. The structure of the nonflexible MIL-100 MOF was retained after encapsulation of the drug as confirmed by x-ray diffraction (XRD). The cytotoxicity of busulfan encapsulated on the nanoparticles of MIL-100(Fe) was determined using human leukemia (CCRF-CEM) and human multiple myeloma (RPMI-8226) cell lines (Horcajada et al. 2010).

Topotecan (Figure 12.3) is a chemical derivative of the cytotoxic drug camptothecin, which is clinically used in the treatment of certain cancers (Möbus et al. 2007).

The encapsulation and controlled release of topotecan on nano-MIL-100(Fe) NMOF was reported (di Nunzio et al. 2014). After microwave-assisted synthesis, the NMOF was washed with ethanol to remove the nonreacted 1,3,5-benzenetricarboxylic (BTC) precursor, but no thermal activation was conducted. The encapsulation of topotecan was performed by impregnation of NMOFs with water solution of topotecan at 22°C and pH $\approx$ 6.7. The topotecan molecule with dimensions $7.6 \times 8.0 \times 13.3$ Å can pass through the hexagonal windows (8.6 Å) to the large cages of the nano-MIL-100(Fe). The UV-Vis diffuse reflectance spectra (DRS) of the encapsulated topotecan showed absorption bands at 396 and 424 nm and a shoulder at 480 nm. These absorption bands were assigned to optical transitions due to the coordination between the Fe_3O center in MIL-100 and CO groups in the adsorbed topotecan molecules,

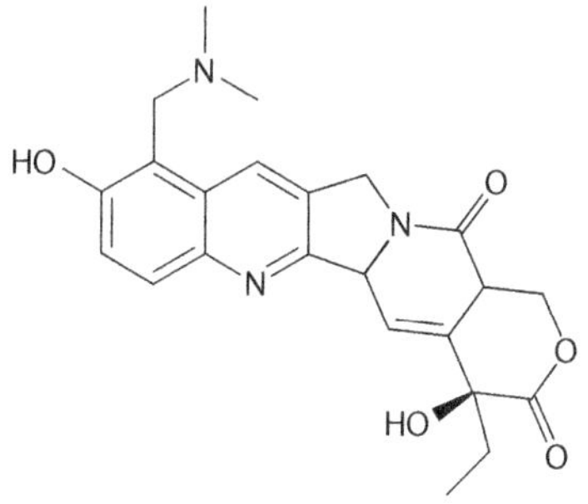

FIGURE 12.3   The structure of the anticancer drug topotecan.

likely due to the formation of the ligand-to-metal charge transfer (LMCT) complex. Experimental data suggested that the monomers of topotecan are accumulated within the same large cage (LC) in the MOF (Figure 12.4). In Figure 12.4, TPT (M) denotes monomers and TPT (D) dimers of topotecan.

For the photo-induced release of encapsulated topotecan to water at pH ≈ 6.7 or to the PBS, illumination with a 60 ps pulsed diode laser (390 nm, 20 MHz) or a 120 fs pulsed Ti:sapphire oscillator (780 nm, 82 MHz) has been used. Photo-induced drug release is believed to work due to weakening of the LMCT complex upon absorption of the photon. Without the photo-induction, the release of topotecan was about five-fold slower. The structure of nano-MIL-100(Fe) was preserved after 3 h of irradiation as determined by powder XRD (di Nunzio et al. 2014).

The efficient novel kinds of drug delivery systems (DDSs) are intended to reme-diate a well-known problem in chemotherapy: high drug doses are needed to com-pensate for the poor biodistribution of administered drugs that exhibit frequent dose-related side effects. Novel MOF-based tumor-targeting DDS was reported by one-pot, solvent-free postsynthetic surface modification of nano-MIL-101(Fe) (Wang et al. 2015, Figure 12.5). First, doxorubicin (DOX) was loaded into the pores of MOF MIL-101–N$_3$(Fe) by impregnation in an aqueous solution. Second, the obtained MIL-101–N$_3$(Fe) composite was further modified with a bicyclononyne-functionalized β-cyclodextrin through azide–alkyne cycloaddition. Next, the surface of the obtained MIL-101 composite was covered with a layer of β-cyclodextrin. Finally, the integrin-targeting peptide-functionalized polymer abbreviated as K(ad)RGDS-PEG1900 was linked to the surface of the MOF. The DOX loaded into this DDS exhibited effective cancer cell inhibition with decreased side effects (Wang et al. 2015).

5-Fluorouracil (5-FU) (Figure 12.6) is a major anticancer drug that has been known at least since 1957 (Bollag 1957). There are few papers in recent years on the encapsulation of 5-FU onto mesoporous MOFs and the controlled release of this drug to bodily fluids, Table 12.2.

The chemical reaction of 5,5′,5″-(1,3,5-triazine-2,4,6-triyl)tris(azanediyl)tri-isophthalate (TATAT) with Zn nitrate in a mixture of dimethylformamide (DMF) and CH$_3$OH yields colorless crystals of the new mesoporous MOF (Sun et al. 2011). This MOF contains two kinds of cages: a hexagonal prism-shaped cage 1.7 nm large and a trigonal prismatic cage 2.1 nm large (Figure 12.7).

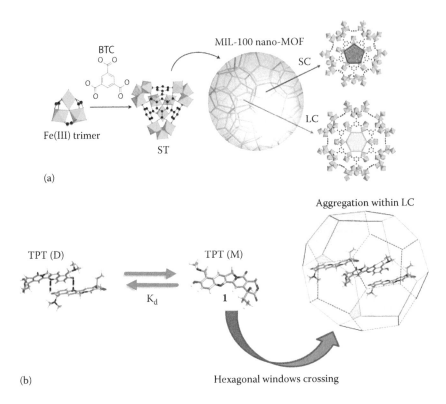

(a)

(b)   Hexagonal windows crossing

FIGURE 12.4 (a) MIL-100 NMOF is generated by the coordination self-assembly of Fe(III) ochtahedral trimers and trimesic acids (BTC) into hybrid supertetrahedra (ST), which assemble giving rise to a rigid micromesoporous zeotypic-like structure. NMOFs possess two types of mesoporous cages: small cages (SCs) accessible through pentagonal windows and large cages (LCs) also delimited by bigger hexagonal openings. (b) Schematic representation of the hypothesized entrapment mechanism of **1** within MIL-100 NMOFs; **1** dimers dissociate, and monomers cross only the hexagonal windows but not the pentagonal ones, so they can be entrapped in only the LC. Entrapped monomers can further aggregate inside the cages by stacking interactions. (Adapted with permission from di Nunzio, M.R., Agostoni, V., Cohen, B., Gref, R., and Douhal, A., A "ship in a bottle" strategy to load a hydrophilic anticancer drug in porous metal organic framework nanoparticles: Efficient encapsulation, matrix stabilization, and photodelivery, *J. Med. Chem.*, 57(2), 411–420. Copyright 2014 American Chemical Society.)

   The windows leading to these cages are 6.3×10.5 Å and 14.3×11.5 Å, respectively. This MOF activated at 100°C overnight was loaded with 5-FU from the methanol solution at room temperature for 3 days. Chemical analysis of the drug-encapsulated MOF gives the content of 5-FU at 33.3 wt.% or 0.5 g/g of MOF. Powder XRD confirmed that the MOF maintains its crystallinity after encapsulation of 5-FU. The Fourier transform infrared (FTIR) spectra confirmed an incorporation of 5-FU within the MOF by the presence of the C–H stretch at 2900 cm^{-1} and –O–C–O– vibrations between 1660 and 1360 cm^{-1}. The shift of the C=O band of the carboxylic group of 5-FU from 1691 to 1723 cm^{-1} indicates the formation of a hydrogen bond

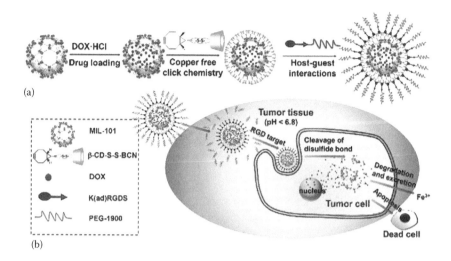

(a)

(b)

FIGURE 12.5 Schematic illustration of (a) drug loading and the postsynthetic modification procedure and (b) the tumor-targeting drug delivery and cancer therapy procedure of the multifunctional MOF-based DDS. (From Wang, X.G., Dong, Z.Y., Cheng, H., Wan, S.S., Chen, W.H., Zou, M.Z., Huo, J.W., Deng, H.X., and Zhang, X.Z., A multifunctional metal-organic framework based tumor targeting drug delivery system for cancer therapy, *Nanoscale*, 7(38), 16061–16070, 2015. Reproduced by permission of The Royal Society of Chemistry.)

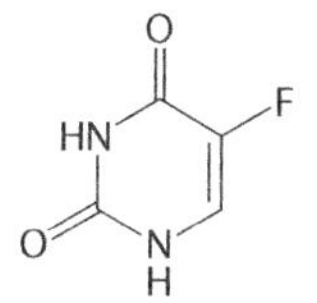

FIGURE 12.6 Molecular structure of 5-fluorouracil drug.

between the CO group of an encapsulated 5-FU and the amino group in the MOF. Drug release was studied by dialysis of the encapsulated MOF in PBS at pH 7.4 and room temperature. There was a steady release of 5-FU to the liquid medium without a "burst effect," and 86.5% of 5-FU was released within a week. Approximately 42% of the loaded amount of 5-FU was released in 8 h, and 43% was released in the two subsequent slower stages. It was speculated that when the drug molecules diffuse near the walls of the pores in the MOF, they interact with the functional groups in the MOF via both hydrogen bonds and π–π interactions, which causes the observed kinetic stages in drug release (Sun et al. 2011).

The $Cu_2(COO)_4$ secondary building unit (SBU) is present in several MOFs including the well-known HKUST-1. The nanocage of metal–organic polyhedron MOP-15 consisting of 12 $Cu_2(COO)_4$ paddlewheel SBUs, 24 $5-NH_2–mBDC$ ligands, and 24 water molecules was used as a precursor for the synthesis of the

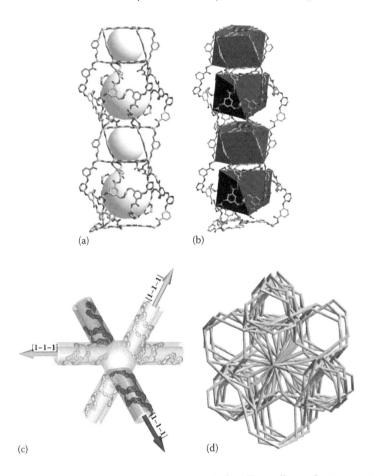

FIGURE 12.7   (a) Two kinds of nanoscale cages in **1 a**. The yellow spheres represent the void inside the cages. (b) Polyhedral presentation of the trigonal and hexagonal prism-shaped nanocages. (c) Representation of the three left-handed double-stranded **3 1** helical chain inter-weaved structure. (d) The four-nodal (4,6)-connected net of **1 a**. Color scheme: C, gray; O, red; N, blue; Zn, green. (From Sun, C.Y., Qin, C., Wang, C.G., Su, Z.M., Wang, S., Wang, X.L., Yang, G.S., Shao, K.Z., Lan, Y.Q., and Wang, E.B.: Chiral nanoporous metal-organic frameworks with high porosity as materials for drug delivery. *Adv. Mater.* 2011. 23(47). 5629–5632. Copyright Wiley-VCH Verlag GmbH & Co. KGaA. Reproduced with permission.)

new MOF (Wang et al. 2011). MOP-15 reacted with 2,2′-bipyridine (bpy) in N,N′-dimethylacetamide (DMA) and ethanol at an elevated temperature to yield the new MOF with the formula $[Cu_{24}(5\text{-}NH_2\text{-}mBDC)_{24}(bpy)_6(H_2O)_{12}]$ DMA_{72} (Figure 12.8).

In the MOF in Figure 12.8, there are two kinds of cages: the microcage (A) and the mesocage (B). The microcage A has a cubo-hemioctahedral shape with a diameter of 16 Å, with the paddlewheel SBUs $Cu_2(COO)_4$. The mesocage B has a truncated octahedral shape with internal diameters of 23.18 Å × 32.63 Å. In the MOF, there are channels with dimensions 10.02 × 7.73 Å through which the adsorbate

TABLE 12.2

Encapsulation and Controlled Desorption/Release of the Anticancer Drug 5-FU from Mesoporous MOFs

MOF	Drug Encapsulation	Host/Guest Bonding	Drug Release	Reference
TATAT + Zn	Activated MOF, methanol	H-bonding	PBS, 25°C, pH 7.4, 1 week	Sun et al. (2011)
Cu derivative of MOP-15	Methanol	—	PBS, 80%, 7.4 h; MOF decomposed	Wang et al. (2011)
Zn–BTC	Activated MOF, methanol	—	PBS, pH 7.4, 42%, 9 h, MOF was stable. Acetate buffer, pH 5.0, 80%, 12 h, MOF was unstable	Wang et al. (2013)

molecule diffuses toward the mesocages. Encapsulation of 5-FU from the methanol solution onto this MOF was conducted for 3 days; the solid was isolated by centrifugation and washed with methanol. The encapsulated amount of 5-FU was found to be 23.76 wt.%; spectroscopic characterization of the molecular form of encapsulated 5-FU was not conducted. The release of 5-FU was conducted in the PBS solution at pH 7.4 by dialyzing against deionized water at 37°C. During the initial stage of the release (7.5 h), about 80% of the encapsulated 5-FU was released, while the second stage of the release was slower. The interactions between the Lewis acid sites in the MOF and the unknown basic sites in 5-FU were assumed to be responsible for the slow release, but no spectroscopic study was performed. The MOF underwent structural degradation during the drug release as was determined by powder XRD (Wang et al. 2011).

The new mesoporous MOF was synthesized from the solution of $Zn(NO_3)_2 \cdot 6H_2O$, H_3BTC precursor of the linker and NaOH in DMA at 85°C (Wang et al. 2013). This MOF denoted compound 2 contains two types of cages (A and B cages) and rhombus-like channels (Figure 12.9).

Cage A has an internal diameter of 14.509 × 13.908 Å. Cage B has Zn_2 SBUs, Zn_3 SBUs, and Zn_4 SBUs linked by 14 BTC ligands and has an internal diameter of 19.727 × 20.751 Å. Before the encapsulation, MOF was activated by heating at 120°C for 24 h in vacuum. Drug encapsulation was performed by suspending the activated MOF in the solution of 5-FU in methanol for 3 days, collecting the solid by centrifugation, washing with methanol, and drying at room temperature. The form of the encapsulated 5-FU (neutral, ionic, hydrogen bonded, etc.) was not determined. The encapsulated amount was determined using UV absorbance in solution at $\lambda_{max} =$ 264 nm. Drug release was performed by dialyzing the 5-FU-loaded MOF in PBS at pH 7.4 and 37°C. About 42% of the encapsulated drug was released in the initial step of 9 h, and ca. 89% was released within the following week. The release was also tested in acetate buffer at pH 5.0, and 80% of the loaded drug was released in 12 h. The accelerated release rate in the acetate buffer compared to that in the PBS is due to the decomposition of the MOF at pH 5.0, as was shown by powder XRD.

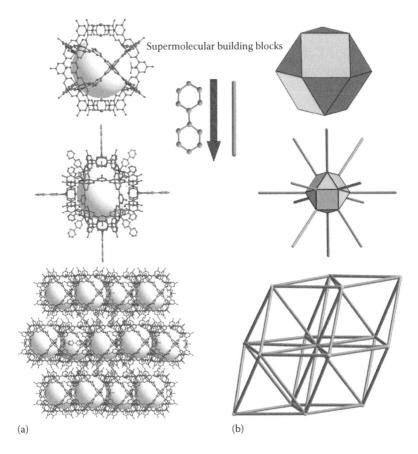

(a)   (b)

FIGURE 12.8   (a) Extended structures based on MOP-15 as precursor with all solvent molecules and hydrogen atoms omitted for clarity. The 5-NH$_2$–mBDC ligand and the Cu atom are highlighted in pink; the bpy ligand is highlighted in green; the yellow spheres indicate the cavity of cuboctahedra. (b) Schematic diagram of appropriate strategy from MOP to MOFs. (Adapted from Wang, H.N., Meng, X., Yang, G.S., Wang, X.L., Shao, K.Z., Su, Z.M., and Wang, C.G., Stepwise assembly of metal-organic framework based on a metal-organic polyhedron precursor for drug delivery, *Chem. Commun.*, 47(25), 7128–7130, 2011. Reproduced by permission of The Royal Society of Chemistry.)

12.4   ENCAPSULATION AND CONTROLLED RELEASE OF ANTIVIRAL DRUGS

Encapsulation and controlled release of antiviral drugs from mesoporous MOFs were first reported in (Horcajada et al. 2010), see Table 12.3. The reported antiviral drugs were anti-HIV (human immunodeficiency virus) drugs that can be considered life-saving, so some moderate toxicity of the Fe-containing MOF sorbents could be permitted. Cidofovir (CDV), L(S)-1-(3-hydroxy-2-phosphonyl-methoxypropyl)cytosine, is an antiviral drug used in the treatment for cytomegalovirus in patients with HIV (Figure 12.10).

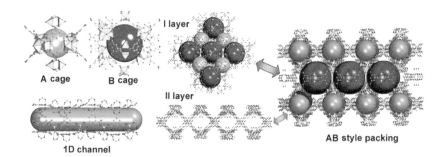

A cage   B cage

I layer

II layer

1D channel

AB style packing

FIGURE 12.9   The schematic representation of compound **2**. (From Wang, H.-N., Yang, G.-S., Wang, X.-L., and Su, Z.-M., pH-induced different crystalline behaviors in extended metal-organic frameworks based on the same reactants, *Dalton Trans.*, 42(18), 6294–6297, 2013. Reproduced by permission of The Royal Society of Chemistry.)

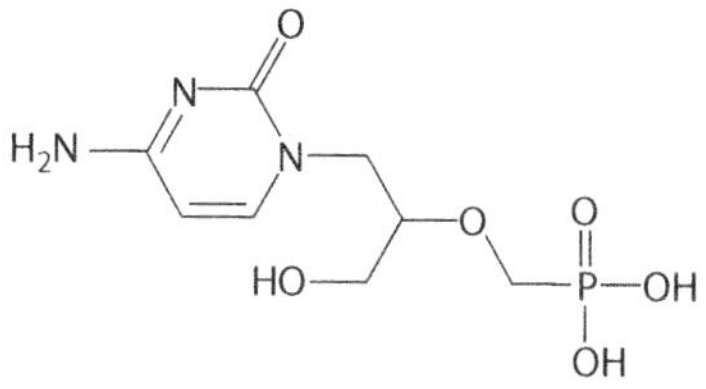

FIGURE 12.10   Structure of the cidofovir antiviral drug.

Prior to adsorption, nano-MIL-100(Fe) was activated by drying overnight at 150°C in air (Horcajada et al. 2010), while nano-MIL-101(Fe)–NH_2 was washed with absolute ethanol for 5 min, centrifuged, and kept wet. Such a precaution is apparently necessary due to the lower temperature and air stability of the MOF with the amino-substituted linker. CDV was encapsulated into nano-MIL-100(Fe) and nano-MIL-101(Fe)–NH_2 by soaking the MOFs into the water solution of CDV at room temperature for 16 h, with some amount of CDV present as the isotope-labeled ^{14}C form. After drug encapsulation, NMOF was removed from the suspension by centrifugation and dried in vacuum for 3 days. On nano-MIL-100(Fe), the loading of CDV is 10.6 wt.% after one incubation and 16.1 wt.% after two successive incubations. On MIL-101(Fe)–NH_2, the encapsulated amount was 22.1% after one incubation and 41.9% after two incubations. The molecular form of the encapsulated CDV and the type of chemical bonds between the encapsulated CDV and the MOF were not determined. The release of CDV was conducted in the PBS at pH 7.4 and room temperature and was quantified by radioactivity measurements of the ^{14}C-labeled CDV. Progressive release of CDV from nano-MIL-100(Fe) occurs without a "burst effect," with the complete release in 5 days (Horcajada et al. 2010). The release kinetics from nano-MIL-101(Fe)–NH_2 was not reported.

Azidothymidine triphosphate (AZT-TP) (Figure 12.11) is a drug frequently used against HIV; AZT-TP inhibits the reverse transcriptase to synthesize DNA of the HIV and thus prevents viral DNA from replication.

TABLE 12.3
Encapsulation and Release of Antiviral Drugs on MIL-100

MOF	Drug	Drug Formula	Encapsulation	Host/Guest Bonding	Release	Reference
Nano-MIL-100(Fe)	Cidofovir (CDV)	Figure 12.10	Water	—	PBS, room temperature, pH 7.4	Horcajada et al. (2010)
Nano-MIL-100(Fe)	Azidothymidine triphosphate (AZT-TP)	Figure 12.11	Water	—	Above	Horcajada et al. (2010)
Nano-MIL-100(Fe)	AZT-TP	Above	Water	Phosphate–Fe(III), LMCT	Water, Tris pH 7.4, PBS pH 7.4	Agostoni et al. (2013)
Nano-MIL-100(Fe)	Azidothymidine (AZT)	Figure 12.12	Water	Nonspecific interactions	Above	Agostoni et al. (2013)

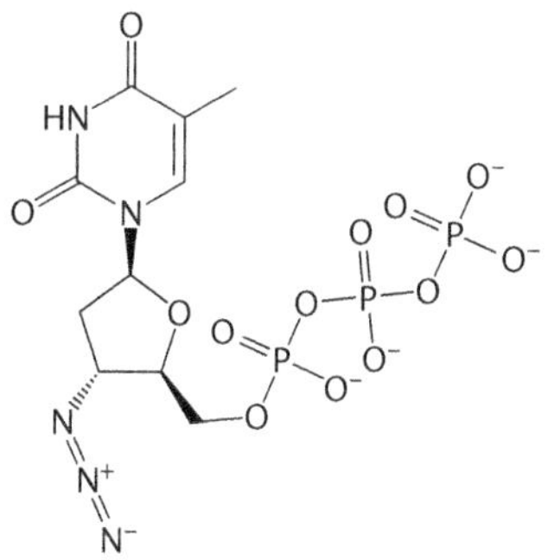

FIGURE 12.11   Azidothymidine triphosphate (AZT-TP).

Encapsulation of AZT-TP was conducted by suspending an activated (150°C overnight) nano-MIL-100(Fe) in a water solution of AZT-TP, with some isotope-labeled ^{3}H AZT-TP present (Horcajada et al. 2010). The encapsulation was performed by stirring the suspension for 16 h at room temperature, and the AZT-TP-loaded MOF was recovered by centrifugation and dried in vacuum for 3 days. For the quantitative analysis, the MOF with the encapsulated drug was decomposed in 5 M HCl at 50°C, and the radioactivity from the ^{3}H AZT-TP in solution was used to calculate the concentration. The encapsulated amount on nano-MIL-100(Fe) was 10.3 wt.% after one incubation and 21.2 wt.% after two incubations. The release of AZT-TP was conducted by suspending the nano-MIL-100(Fe)-encapsulated AZT-TP in PBS at pH 7.4, and the released amount was determined by the radioactivity measurement. Progressive release of AZT-TP proceeds without the "burst effect," with complete release in 3 days. During 3 days of release, only ca. 10% of the total amount of nano-MIL-100(Fe) was decomposed. It was suggested that the AZT-TP drug was adsorbed within the mesopores of the nano-MIL-100(Fe). The encapsulation of AZT-TP on nano-MIL-100(Fe)–NH$_2$ is as high as 25.4 wt.% after one incubation and 42.0 wt.% after two incubations. The release kinetics of AZT-TP from MIL-100(Fe)–NH$_2$ has not been reported (Horcajada et al. 2010). Later, the same group reported the mechanistic study of the encapsulation of the neutral molecule AZT (Figure 12.12) versus anionic AZT-TP on nano-MIL-100(Fe) by adsorption in water at room temperature (Agostoni et al. 2013).

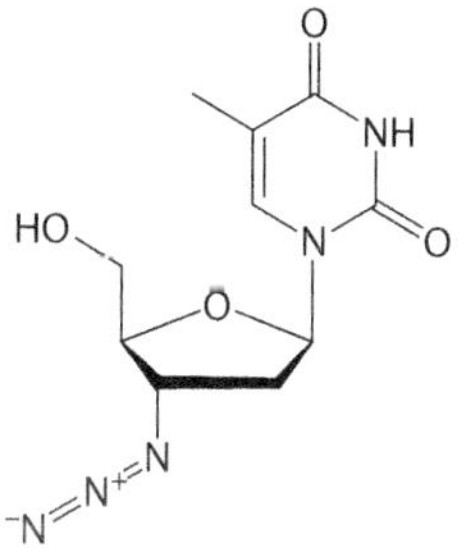

FIGURE 12.12   Azidothymidine (AZT).

The supramolecular structure of nano-MIL-100(Fe) remained unchanged after encapsulation of AZT-TP, as determined by powder XRD (Agostoni et al. 2013). According to Monte Carlo simulations, both AZT and AZT-TP molecules are located in the mesocages of nano-MIL-100(Fe). The phosphate group of AZT-TP is close to the Fe(III) coordinatively unsaturated site (CUS) of nano-MIL-100(Fe). The UV-Vis diffuse reflectance spectrum (UV-Vis DRS) of the AZT-TP/nano-MIL-100(Fe) showed the new absorption band at the larger wavelength, compared to those of both nano-MIL-100(Fe) and AZT-TP. This new band was attributed to the LMCT transition due to coordination between the phosphate group of adsorbed AZT-TP and Fe(III) CUS. The release of AZT-TP was concluded to be not due to decomposition of the MIL-100(Fe) drug carrier in the buffer, but due to the substitution of AZT-TP coordinated with Fe(III) with phosphate anion from the PBS (Agostoni et al. 2013).

Contrary to the AZT-TP, the total amount of AZT adsorbed on nano-MIL-100(Fe) was quite small: only 2.2 wt.% (Agostoni et al. 2013). The molecular size of AZT is smaller than that of AZT-TP; therefore, the much smaller encapsulated amount of AZT was thought to be not due to the size of the molecule. Instead, the difference between AZT and AZT-TP was thought to be due to the strong bonds between the phosphate group of AZT-TP and Fe(III) in nano-MIL-100(Fe). The UV-Vis diffuse reflectance (UV-Vis DRS) spectrum of the AZT-encapsulated nano-MIL-100(Fe) is the sum of the contributions from the AZT adsorbate and nano-MIL-100(Fe); therefore, no coordination bond was formed. The weak interactions between nitrogen atoms in encapsulated AZT and phenyl rings and oxygen atoms of COO group of the BTC linker in nano-MIL-100(Fe) were found by quantum chemical simulations. AZT was released with the "burst effect," which was another indication of weak bonding of the encapsulated form of AZT to the adsorption sites in nano-MIL-100(Fe).

12.5 ENCAPSULATION AND RELEASE OF MISCELLANEOUS BIOACTIVE COMPOUNDS

For the drugs intended to treat noncritical conditions, the use of engineered nanoparticles as drug carriers needs to be explicitly justified, given some cytotoxicity of many metals and linkers present in the MOFs. In this case, the risk associated with the unknown toxicity of the engineered encapsulated form of the drug can overweigh the therapeutic benefits of the administered drug.

A good example is encapsulation of the common over-the-counter (OTC) drug ibuprofen (IBU) onto MOFs (Horcajada et al. 2006), Figure 12.13 and Table 12.4.

IBU was encapsulated by adsorption onto the activated MIL-100(Cr) and MIL-101(Cr) from a solution in hexane. MIL-100(Cr) and MIL-101(Cr) MOFs are considered relatively toxic due to the presence of Cr(III). For MIL-100(Cr), the maximum adsorbed amount of IBU was 0.347 g/g sorbent, while it was as high as 1.376 g/g for MIL-101(Cr). The IBU molecule has dimensions of ca. 5×10 Å. In MIL-100(Cr), the diameter of the window leading to the mesocavity is 9.7 Å, while in MIL-101(Cr) the diameter of the window is 12.6 Å. Therefore, it is easier for the IBU molecule to reach the mesocavity in MIL-101(Cr) than the mesocavity in MIL-100(Cr), which explains

TABLE 12.4

Encapsulation and Controlled Release of Miscellaneous Bioactive Compounds on MIL-100

MOF	Biologically Active Organic Compound	Use	Formula	Reference
MIL-100(Cr)	Ibuprofen (IBU)	NSAID	Figure 12.13	Horcajada et al. (2006)
Nano-MIL-100(Fe)	Oxybenzone, benzophenone-3	UVA sunscreen	Figure 12.14	Horcajada et al. (2010)
Nano-MIL-100(Fe)	Sulisobenzone	UVA sunscreen	Figure 12.15	Horcajada et al. (2010)
Nano-MIL-100(Fe)	Caffeine	Stimulant	Figure 12.16	Horcajada et al. (2010)
Nano-MIL-100(Fe)	Urea	Ingredient in cosmetics	$(NH_2)_2C=O$	Horcajada et al. (2010)
Nano-MIL-100(Fe)	Caffeine	Stimulant	Figure 12.16	Cunha et al. (2013)

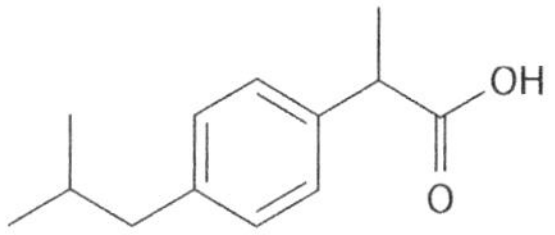

FIGURE 12.13   Structure of ibuprofene (IBU).

the different maximum encapsulated amounts (Horcajada et al. 2006). It is important to consider that both MIL-100(Cr) and MIL-101(Cr) are "rigid" MOFs and their structural parameters do not change upon adsorption of the "guest" drug molecule. Therefore, the structural parameters of MIL-100 and MIL-101 can be used to predict the adsorption of a molecule of given size. On the other hand, in "flexible" MOFs such as MIL-53, the "breathing" effects, that is, the changes in the structural parameters during adsorption (Loiseau et al. 2004), would need to be taken into account. The ^{13}C nuclear magnetic resonance (NMR) spectra showed that the α-alpha carbon atom (the one close to CO group of IBU) is strongly shifted upon encapsulation of IBU (Horcajada et al. 2006). The solid-state ^1H NMR of the "free" IBU showed the protonated (acid) form, as expected, at $\delta = 14$ ppm, while the ^1H NMR spectra of an encapsulated IBU showed the disappearance of the peak of the protonated form; therefore, IBU exists as an anion in its adsorbed form. The countercation that would charge-compensate the encapsulated anion of IBU was not determined. The release kinetics from MIL 100(Cr) to the simulated body fluid (SBF) at 37°C in the first 2 h can be fitted with zero-order kinetics, and complete release was achieved in 3 days. It took 6 days for complete release of IBU from MIL-101(Cr). The longer release of IBU from MIL-101(Cr) versus MIL-100(Cr) was speculated to be due to the higher proportion of aromatic rings of the linker of MIL-101 versus MIL-100 to the Cr(III) CUS, with aromatic rings serving as tentative adsorption sites for IBU molecule.

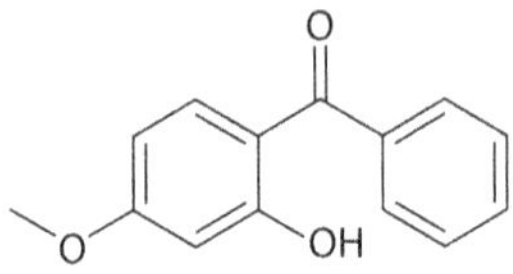

FIGURE 12.14   Oxybenzone, aka benzophenone-3.

Oxybenzone, aka benzophenone-3 (Figure 12.14), is present in sunscreens, nail polish, hairspray, cosmetics, and fragrances where it acts as a photostabilizer.

The amount of oxybenzone encapsulated onto nano-MIL-100(Fe) reached only 1.5 wt.% (Horcajada et al. 2010), and the nature of bonding to the MOF and release kinetics were not determined.

Sulisobenzone, aka benzophenone-4 (Figure 12.15), is used in sunscreens, and it is the derivative of oxybenzone with the polar SO_3H group.

The encapsulation of polar sulisobenzone on nano-MIL-100(Fe) was reported in the same paper (Horcajada et al. 2010) and was compared to that of oxybenzone. Nano-MIL-100(Fe) was activated at 100°C for 12 h and suspended in aqueous solution of the compound of interest in water under stirring for 3 days at room temperature, recovered by centrifugation, and dried in air. The encapsulated amount was determined by elemental analysis and thermogravimetric analysis (TGA). The maximum loading of polar sulisobenzone at 15.2 wt.% in water is 10 times higher than that of the nonpolar oxybenzone (Horcajada et al. 2010). The higher adsorbed amount of the polar aromatic organic compounds versus the nonpolar compounds of otherwise similar structure has been noted in nonwater solvents as well (e.g., Henschel et al. 2011).

Alkaloid caffeine (Figure 12.16) is a stimulant of the nervous system and a lipo-reductor in cosmetics.

Encapsulation of caffeine is a challenge for the cosmetic industry, since caffeine tends to crystallize, which leads to poor loadings (<5 wt.%) and uncontrolled release. Nano-MIL-100(Fe) activated at 100°C was suspended in an aqueous solution of

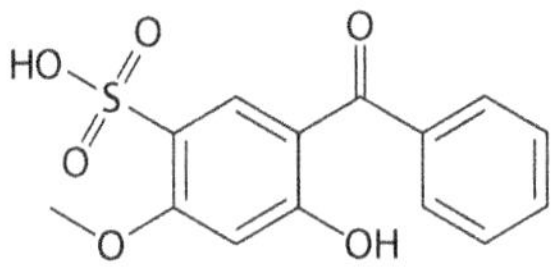

FIGURE 12.15   Sulisobenzone, aka benzophenone-4.

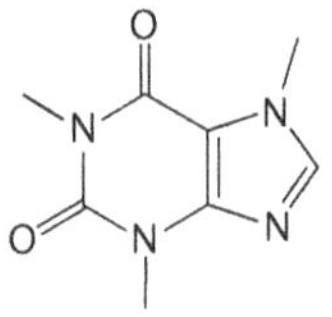

FIGURE 12.16   Caffeine.

caffeine under stirring for 3 days, recovered by centrifugation, and dried in air. The encapsulated amount of caffeine was 24.2% (Horcajada et al. 2010); interestingly, this amount is about the same as was found for the encapsulation on *microporous* MIL-53(Fe) at 23.1%.

Another systematic study of driving forces for encapsulation and release of caffeine from nano-MIL-100(Fe) was reported (Cunha et al. 2013). The nano-MIL-100(Fe) was activated at 150°C for 16 h, and caffeine was encapsulated by suspending the activated nano-MIL-100(Fe) in an aqueous solution, under stirring at room temperature for 72 h. The caffeine-loaded nano-MOF was removed by filtration and dried at room temperature. The encapsulation of caffeine reached 50 wt.% of the amount of the MOF that was approaching the maximum theoretical loading at 65.8 wt.%, assuming full accessibility of *both* the small pentagonal windows (4.7 × 5.5 Å) and the large windows of 8.6 Å to the caffeine adsorbate. However, the molecular size of caffeine at 7.6 × 6.1 Å allowed its adsorption in the larger windows only with the predicted loading at 46.4 wt.%, so the predicted loading was in good agreement with experimentally measured loading at 50 wt.%. The crystalline structure of nano-MIL-100(Fe) did not change after encapsulation as was shown by powder XRD, and no recrystallized caffeine was observed within the MOF host material. The encapsulated caffeine showed the vibration of the ν(C=O) group at 1700 cm^{-1} by FTIR, while "free" caffeine had this vibration at 1691 cm^{-1}. The shift to the higher frequency suggests weak interactions between the nitrogen atom (located between two C=O groups in caffeine molecule) and the functional groups of the MOF as was shown by using the model compounds with the C=O group, ethyleneurea and propyleneurea. Which exactly functional groups in the nano-MIL-100(Fe) sorbent interact with an encapsulated caffeine molecule was not determined by experiments. It should be noted that the details of bonding are difficult to determine for an adsorbate with a rather complex chemical structure. The release of caffeine at 37°C into PBS at pH = 7.4 occurs fully in just 0.5 h, while the release into pure water at pH = 6.3 at 37°C occurs as slowly as within 48 h. Fast and complete caffeine release into the PBS solution is accompanied by a collapse of the framework of nano-MIL-100(Fe). The destruction of the framework is due to the replacement of the carboxylate BTC linkers by phosphate groups, as was shown by FTIR via the presence of the ν(PO$_4^{3-}$) spectral bands in nano-MIL-100(Fe) after the release of caffeine (Cunha et al. 2013).

Given that many MOFs contemplated as drug nano-carriers are chemically destroyed in cellular media, the drugs intended for "noncritical" conditions, for example, IBU (Horcajada et al. 2006) and caffeine (Cunha et al. 2013), would require MOF carriers with very low toxicity. The best MOFs for this purpose are those containing nontoxic Al, Fe, or Ti metals and the linkers made of the biochemically "benign" nucleotide bases or amino acids. These MOFs have been termed biocompatible MOFs, or BIO-MOFs (e.g., An et al. 2009). BIO-MOFs can be prepared in the form of nanocolloids, that is, as NMOFs. Research with mesoporous nano-BIO-MOFs for drug carrier applications is currently under way, and many interesting findings are anticipated.

The rate of adsorption of vitamin B12 in water on mesoporous TbMOF-100 was studied at room temperature and at 37°C in 1-hour intervals (Valencia 2013).

After 19.3 h of adsorption, the concentration of vitamin B12 in the solution decreased from 2.23 to 1.3 mM. The initial rate of adsorption was found to be 0.088 mM/h, and the coefficient of diffusion was found to be 66 nm^2/h using Fick's law. The value of the coefficient of diffusion indicates that the adsorption of vitamin B12 includes the step of diffusion inside the cages of mesoporous sorbent TbMOF-100.

REFERENCES

Agostoni, V., T. Chalati, P. Horcajada, H. Willaime, R. Anand, N. Semiramoth, T. Baati et al. 2013. Towards an improved anti-HIV activity of NRTI via metal-organic frameworks nanoparticles. *Advanced Healthcare Materials* 2(12):1630–1637.

An, J. Y., S. J. Geib, and N. L. Rosi. 2009. Cation-triggered drug release from a porous zinc-adeninate metal-organic framework. *Journal of the American Chemical Society* 131(24):8376–8377.

Aubel-Sadron, G. and D. Londos-Gagliardi. 1984. Daunorubicin and doxorubicin, anthracycline antibiotics, a physicochemical and biological review. *Biochimie* 66(5):333–352.

Bollag, W. 1957. Experimental studies on 5-fluorouracil, a cytostatic agent. *Schweizerische medizinische Wochenschrift* 87(26):817–820.

Cunha, D., M. Ben Yahia, S. Hall, S. R. Miller, H. Chevreau, E. Elkaim, G. Maurin, P. Horcajada, and C. Serre. 2013. Rationale of drug encapsulation and release from biocompatible porous metal-organic frameworks. *Chemistry of Materials* 25(14):2767–2776.

di Nunzio, M. R., V. Agostoni, B. Cohen, R. Gref, and A. Douhal. 2014. A "ship in a bottle" strategy to load a hydrophilic anticancer drug in porous metal organic framework nanoparticles: Efficient encapsulation, matrix stabilization, and photodelivery. *Journal of Medicinal Chemistry* 57(2):411–420.

Goekkurt, E., J. Stoehlmacher, C. Stueber, C. Wolschke, T. Eiermann, S. Iacobelli, A. R. Zander, G. Ehninger, and N. Kröger. 2007. Pharmacogenetic analysis of liver toxicity after busulfan/cyclophosphamide-based allogeneic hematopoietic stem cell transplantation. *Anticancer Research* 27(6C):4377–4380.

Henschel, A., I. Senkovska, and S. Kaskel. 2011. Liquid-phase adsorption on metal-organic frameworks. *Adsorption* 17(1):219–226.

Horcajada, P., T. Chalati, C. Serre, B. Gillet, C. Sebrie, T. Baati, J. F. Eubank et al. 2010. Porous metal-organic-framework nanoscale carriers as a potential platform for drug delivery and imaging. *Nature Materials* 9(2):172–178.

Horcajada, P., C. Serre, M. Vallet Regí, M. Sebban, F. Taulelle, and G. Férey. 2006. Metal–organic frameworks as efficient materials for drug delivery. *Angewandte Chemie International Edition* 45(36):5974–5978.

Li, Z. H., Z. Liu, M. L. Yin, X. J. Yang, Q. H. Yuan, J. S. Ren, and X. G. Qu. 2012. Aptamer-capped multifunctional mesoporous strontium hydroxyapatite nanovehicle for cancer-cell-responsive drug delivery and imaging. *Biomacromolecules* 13(12):4257–4263.

Loiseau, T., C. Serre, C. Huguenard, G. Fink, F. Taulelle, M. Henry, T. Bataille, and G. Ferey. 2004. A rationale for the large breathing of the porous aluminum terephthalate (MIL-53) upon hydration. *Chemistry—A European Journal* 10(6):1373–1382.

Möbus, V., D. G. Kieback, and S. K. Kaubitzsch. 2007. Duration of chemotherapy with topotecan influences survival in recurrent ovarian cancer: A meta-analysis. *Anticancer Research* 27(3B):1581–1587.

Samuel, J., P. Elamanchili, C. Chong, C. Lutsiak, K. Newman, M. Diwan, and G. S. Kwon. 2003. Biodegradable nanoparticles for targeted delivery of therapeutic vaccines to dendritic cells. *FASEB Journal* 17(7):C332.

Sun, C. Y., C. Qin, C. G. Wang, Z. M. Su, S. Wang, X. L. Wang, G. S. Yang, K. Z. Shao, Y. Q. Lan, and E. B. Wang. 2011. Chiral nanoporous metal-organic frameworks with high porosity as materials for drug delivery. *Advanced Materials* 23(47):5629–5632.

Valencia, V. 2013. Diffusion of vitamin B12 across a mesoporous metal organic framework. *Undergraduate Journal of Mathematical Modeling: One + Two* 5(1):Article 3.

Wang, H.-N., X. Meng, G. S. Yang, X. L. Wang, K. Z. Shao, Z. M. Su, and C. G. Wang. 2011. Stepwise assembly of metal-organic framework based on a metal-organic polyhedron precursor for drug delivery. *Chemical Communications* 47(25):7128–7130.

Wang, H. N., G. S. Yang, X.-L. Wang, and Z.-M. Su. 2013. pH-induced different crystalline behaviors in extended metal-organic frameworks based on the same reactants. *Dalton Transactions* 42(18):6294–6297.

Wang, X. G., Z. Y. Dong, H. Cheng, S. S. Wan, W. H. Chen, M. Z. Zou, J. W. Huo, H. X. Deng, and X. Z. Zhang. 2015. A multifunctional metal-organic framework based tumor targeting drug delivery system for cancer therapy. *Nanoscale* 7(38):16061–16070.

White, J. M. L., A. M. Powell, K. Brady, and R. Russell-Jones. 2003. Severe generalized argyria secondary to ingestion of colloidal silver protein. *Clinical and Experimental Dermatology* 28(3):254–256.

Yokochi, T. and K. D. Robertson. 2004. Doxorubicin inhibits DNMT1, resulting in conditional apoptosis. *Molecular Pharmacology* 66(6):1415–1420.

Zhan, Q., B. Y. Shen, X. X. Deng, H. Chen, J. B. Jin, X. Zhang, C. H. Peng, and H. W. Li. 2013. Drug-eluting scaffold to deliver chemotherapeutic medication for management of pancreatic cancer after surgery. *International Journal of Nanomedicine* 8:2465–2472.

Zhang, T. and C. L. Sentman. 2011. Cancer immunotherapy using a bispecific NK receptor fusion protein that engages both T cells and tumor cells. *Cancer Research* 71(6):2066–2076.

13 Research on Mesoporous MOFs for Industrial Applications

Many advantages of mesoporous metal–organic frameworks (MOFs) for adsorption-based applications such as separation, purification, chromatography, catalysis, and so on are expected to contribute to the development of respective emerging industrial technologies. However, at least one of the major limitations of the wide-scale research and applications of mesoporous MOFs is their limited availability. Indeed, the most typical lab-scale chemical synthesis using the autoclave or even microwave radiation yields several hundred milligrams of the given MOF per procedure. This may be sufficient for the adsorption experiments conducted in the gas phase but is usually not enough to conduct several adsorption experiments in solution, especially when the sorbent has a limited regeneration capability. In addition, the usual problem is variable "quality" of the MOF obtained from one "synthetic batch" to another, as judged by the combination of chemical analysis and structural and spectroscopic characterization. Table 13.1 shows published data on large-scale synthesis of mesoporous MOFs.

Basolite F300 is the most important commercially available mesoporous MOF, and its industrial-scale manufacturing has been described by researchers from Badische Anilin und Soda Fabrik (BASF) (Czaja et al. 2009). An important parameter in the upscale of chemical synthesis is the space–time yield, in addition to the "usual" yield as a measure of success of chemical synthesis. An example of the development of large-scale synthesis of mesoporous MOFs is provided in Seo et al. (2012). The use of toxic and corrosive hydrofluoric acid is avoided. This method, combined with the two purification steps (solvent extraction and chemical treatment with NH_4F), leads to a highly porous material that is obtained with a large space–time yield at >1700 kg/m^3 day. The obtained MOF material had major physicochemical properties similar to those of MIL-100(Fe) prepared by "conventional" lab-scale synthesis using HF.

Out of the MIL-101 family, MIL-101(Cr) is one of the most important mesoporous MOFs for adsorption and catalysis in solution. For large-scale synthesis, it is essential to eliminate the use of hydrofluoric acid, in order to make the process safer and also more "green." Specifically, hydrofluoric acid easily penetrates tissue; thus, life-threatening poisoning can occur readily through exposure of skin or eyes, and even more readily when inhaled or swallowed. Therefore, hydrofluoric acid must be handled with extreme care, protective equipment should be used properly, and stringent safety precautions must be met beyond those normally required for the other mineral acids, to avoid severe chemical burns (Yamashita et al. 2001).

TABLE 13.1

Publications on Large-Scale Synthesis of Mesoporous MOFs

MOF	Method of Synthesis	Space–Time Yield (kg/m³ Day)	Total Surface Area (m²/g)	Reference
Basolite F300	Not disclosed; electrochemical?	20	1300–1600	Czaja et al. 2009
MIL-100(Fe)	Hydrothermal	>1700	Up to 2000	Seo et al. (2012)
MIL-101(Cr)	Hydrothermal in autoclave	—	Up to 4000	Zhao et al. (2015)

In addition, elimination of toxic HF from the synthesis corresponds to "Less Hazardous Chemical Syntheses" as Principle #3 of green chemistry (Anastas and Warner 1998).

Specifically, to replace HF in large-scale synthesis of MIL-101(Cr), the selection of chemical "modifiers" has been systematically tested in recent work (Zhao et al. 2015). The list of investigated modifiers that are acids includes nitric acid, acetic acid, trifluoroacetic acid, sulfuric acid, hydrochloric acid, phenylphosphonic acid, benzoic acid, formic acid, fumaric acid, citric acid, and succinic acid. In addition, the effects of inorganic base NaOH versus organic base tetramethylammonium hydroxide (TMAOH) have been investigated. The target properties of MIL-101(Cr) included high yield, Brunauer–Emmett–Teller (BET) surface area, Langmuir surface area, pore volume, and low temperature of synthesis. Characterization of the products was conducted by elemental analysis, powder x-ray diffraction (XRD), and nitrogen sorption. Out of tested modifiers as alternatives to HF, nitric and acetic acids were the best. Specifically, the advantage of HNO_3 is that there is about 30% increase in the yield compared to conventional MIL-101(Cr) synthesized using HF. The second advantage is the rather high BET surface area at >3100 m²/g, which is only ca. ~10% below that of "conventional" MIL-101(Cr) synthesized with HF. Nitric acid has also been demonstrated to be effective in the large scale synthesis of MIL-101(Cr) at the large lab scale 3 L reaction volume (Zhao et al. 2015).

Amongst experimental parameters of chemical synthesis conducted in heterogeneous media that are hard to reproduce is stirring of reaction mixture. In the synthesis of MIL-101, a hermetically sealed autoclave is frequently used. In the reported small lab-scale synthesis of MIL-101, no stirring was conducted (e.g., Ramos-Fernandez et al. 2011, Yang and Yan 2011, Wang et al. 2013). In the cited report (Zhao et al. 2015), stirring of the reaction mixture was found to be unnecessary to form MIL-101(Cr) with high surface area. Moreover, stirring leads to a MIL-101 material with lower total BET surface area (Figure 13.1).

In addition, the high temperature required for chemical synthesis is another concern. In this research (Zhao et al. 2015), the reaction temperature could be lowered to 160°C by using acetic acid; however, the yield decreased to 50% with the rather acceptable BET surface area of 2700–2800 m²/g. The second advantage of the lower temperature of synthesis is obtaining one-pot prepared composite materials

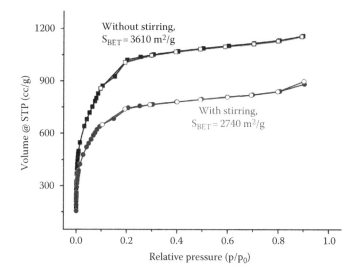

FIGURE 13.1 Nitrogen sorption isotherms and their corresponding BET surface areas for non/stirring experiments of large-scale preparations of MIL-101(Cr); filled symbols are for adsorption and empty symbols for desorption. The synthesis and treatment of both the products followed the general procedure. The BET surface areas were calculated in the pressure range $0.05 < p/p_0 < 0.2$ from N_2 sorption isotherms at 77 K with an estimated standard deviation of ± 50 m^2/g. (From Zhao, T., Jeremias, F., Boldog, I., Nguyen, B., Henninger, S.K., and Janiak, C., High-yield, fluoride-free, and large-scale synthesis of MIL-101(Cr), *Dalton Trans.*, 44(38), 16791–16801, 2015. Adapted by permission of The Royal Society of Chemistry.)

with MIL-101 as the structural host, so that the high temperature of synthesis does not cause decomposition of the "guest" component of the composite, or chemical precursor(s) of "guest" compound(s). Finally, by using HNO_3 under the optimized conditions, MIL-101(Cr) was synthesized in >100 g quantities with the yields at about 70% and the BET surface areas close to 4000 m^2/g. The use of NH_4F at the last step of synthesis of this MIL-101(Cr) was avoided.

One can mention that this large-scale method of synthesis utilizes an autoclave whose volume apparently becomes a limiting factor, once further upscale of production of the MOF is needed. In addition, dimethylformamide (DMF) is used in the MOF purification step for reported large-scale synthesis of MIL-101(Cr) (Zhao et al. 2015) to dissolve the excess of terephthalic acid used as a precursor of the linker. Common organic solvents have been assessed based on their effects on the environment, health, and safety and classified as being "green" (preferred ones), "yellow" (usable), and "red" (undesirable) (Alfonsi et al. 2008). On scale of being "green," DMF is a "red" solvent whose use is not desired in chemical manufacturing. A search for more "green" organic solvents than DMF for the purification of mesoporous and microporous MOFs is needed. Ultimately, the most "green" solvents are water and supercritical carbon dioxide, which should be investigated for the purification of MOFs in the emerging large-scale synthesis.

TABLE 13.2
Manufacturers and Vendors of Mesoporous MOFs

Manufacturer	MOF	MOF Properties	Available Amounts	Reference
BASF; vendor Sigma Aldrich	Basolite F300	$1300-1600$ m²/g	$10-500$ g	Czaja et al. (2009)
Strem Chemicals	KRICT F100	Fluorine-free MIL-100(Fe), 1950 m²/g	15 kg	—
Technische Universität Dresden (TUD)	MIL-101(Cr)(F)	$2600-3100$ m²/g	1 g	Website of TUD
TUD	MIL-101(Fe)–NH₂	$1800-2200$ m²/g	$1-10$ g	Above
TUD	FeBTC spheres	$800-1300$ m²/g, particle size 0.5–2 mm	10 g	Above
TUD	FeBTC–Xerogel	$700-1000$ m²/g	$1-1000$ g	Above

Table 13.2 shows the list of available (March 2016) manufacturers and vendors of mesoporous MOFs. KRICT denotes Korea Research Institute of Chemical Technology.

Materials Center at the Technische Universität Dresden (TUD) supplies several mesoporous MOFs as on the website http://www.metal-organic-frameworks.eu. These MOFs are synthesized and characterized by Stefan Kaskel's group.

Amongst major factors that limit industrial applications of mesoporous MOFs, one could mention their rather high current prices. The major component of the price of the MOF is the price of the organic precursor that forms the linker unit. For the MIL-101 family, the precursor is 1,4-benzenedicarboxylic (terephthalic) acid, which is a commodity chemical with several million tonnes produced annually from p-xylene of petroleum origin see e.g. (Cheng et al. 2006). Terephthalic acid is widely used in industrial organic synthesis as a precursor of the polyethylene terephthalate ester (PETE) polymer, which is widely used to make consumer goods such as clothes, plastic bottles, etc. One can expect a significant price reduction of MOFs made on a large scale from relatively inexpensive precursors such as terephthalic acid.

About three decades have been spent on fundamental research on MOFs concentrating on their synthesis, learning about their properties and studying their adsorption-based applications including purification of fuels, catalysis, environmental cleaning, and health care. MOFs and particularly mesoporous MOFs have been and currently remain a "hot topic" in the academic community. The usability of MOFs in emerging industrial applications is important at this stage; indeed, the extent of their "true" usability and concomitant limitations and challenges has been discussed in recent review papers (e.g., Silva et al. 2015). The cited review covers the range from the very basic concepts of MOF engineering and synthesis through their industrial production on a large scale and applications for the needs of contemporary society (Silva et al. 2015). Some patent literature on the topic has been covered therein, but there was no study specifically on mesoporous MOFs.

In addition to the investigators from the universities wishing to patent their research, there are private companies that conduct research on available patent literature and publish reports on "patent landscapes." The benchmark mesoporous MOFs MIL-100 and MIL-101 are from the Institut Lavoisier in France, and their names denote "Materials of Institut Lavoisier". The company France Innovation Scientifique et Transfert (FIST SA) working in the field of intellectual property (IP) has developed a significant expertise in MOFs. This company has produced the so-called "IP Competitive Landscape" report, which includes all the patents related to MOFs filed since 1990, but does not specifically differentiate mesoporous MOFs from the other kinds of MOFs.

We provide the most typical kinds of patents on the adsorption-based applications of mesoporous MOFs in Table 13.3.

In a recent review paper, academic research on the antimicrobial properties of MOFs was summarized (Wyszogrodzka et al. 2016); there are only few pertinent reports on mesoporous MOFs. Nevertheless, emerging applications of mesoporous MOFs in biomedicine are of high interest for patenting. Antimicrobial MOFs have been patented by researchers from the United Kingdom (Morris 2013). Based on recent research on the production of NO in physiological fluid using microporous copper(II) benzene-1,3,5-tricarboxylate MOF (Neufeld et al. 2015), the use of its mesoporous varieties has been patented (Reynolds and Reynolds 2014). Applications of mesoporous MOFs as composite biocatalysts carrying enzyme microperoxidase-11 have been patented (Ma et al. 2015). Since research on using mesoporous MOFs for biomedical applications is in its infancy, patent literature is quite limited, but the progress in this field is quick with commercial opportunities arising.

TABLE 13.3
Patents on Using Mesoporous MOFs for Adsorption in Solution

MOFs	Patented	Publication Number	Filing Date	Owner	Reference
Various MOFs; application implies mesoporosity	Antibiotic molecules entrapped by the MOF	EP2603076 A2	August 8, 2011	University of St Andrews, United Kingdom	Morris (2013)
$Cu_3(1,3,5,-BTC)_2$, pore size 5–500 Å	Production of NO in physiological fluid	US8771756 B2	December 28, 2010	Colorado State University, United States	Reynolds and Reynolds (2014)
Tb–mesoMOF	Synthesis, biocatalysis with *micro-peroxidase*-11 (MP-11) enzyme in solution	US201500 87044 A1	June 15, 2012	University of South Florida, United States	Ma et al. (2015)

REFERENCES

Alfonsi, K., J. Colberg, P. J. Dunn, T. Fevig, S. Jennings, T. A. Johnson, H. P. Kleine et al. 2008. Green chemistry tools to influence a medicinal chemistry and research chemistry based organisation. *Green Chemistry* 10(1):31–36.

Anastas, P. T. and J. C. Warner. 1998. *Green Chemistry: Theory and Practice.* Oxford University Press, New York.

Cheng, Y., X. Li, L. Wang, and Q. Wang. 2006. Optimum ratio of Co/Mn in the liquid-phase catalytic oxidation of p-xylene to terephthalic acid. *Industrial & Engineering Chemistry Research* 45(12):4156–4162.

Czaja, A. U., N. Trukhan, and U. Muller. 2009. Industrial applications of metal-organic frameworks. *Chemical Society Reviews* 38(5):1284–1293.

Ma, S., L. J. Ming, Y. Chen, and V. Lykourinou. 2015. Polyhedral cage-containing mesoporous metal-organic frameworks as platform for biocatalysis, methods of making these frameworks, and methods of using these frameworks. Patent US20150087044 A1.

Morris, R. E. 2013. Anti-microbial metal organic framework. Patent EP2603076 A2.

Neufeld, M. J., J. L. Harding, and M. M. Reynolds. 2015. Immobilization of metal–organic framework copper(II) benzene-1,3,5-tricarboxylate (CuBTC) onto cotton fabric as a nitric oxide release catalyst. *ACS Applied Materials & Interfaces* 7(48):26742–26750.

Ramos-Fernandez, E. V., M. Garcia-Domingos, J. Juan-Alcañiz, J. Gascon, and F. Kapteijn. 2011. MOFs meet monoliths: Hierarchical structuring metal organic framework catalysts. *Applied Catalysis A: General* 391(1–2):261–267.

Reynolds, M. M. and B. P. Reynolds. 2014. Biocompatible materials for medical devices. Patent US8771756 B2.

Seo, Y.-K., J. W. Yoon, J. S. Lee, U. H. Lee, Y. K. Hwang, C.-H. Jun, P. Horcajada, C. Serre, and J.-S. Chang. 2012. Large scale fluorine-free synthesis of hierarchically porous iron(III) trimesate MIL-100(Fe) with a zeolite MTN topology. *Microporous and Mesoporous Materials* 157:137–145.

Silva, P., S. M. F. Vilela, J. P. C. Tome, and F. A. Almeida Paz. 2015. Multifunctional metal-organic frameworks: From academia to industrial applications. *Chemical Society Reviews* 44(19):6774–6803.

Wang, S., L. Bromberg, H. Schreuder-Gibson, and T. A. Hatton. 2013. Organophophorous ester degradation by chromium(III) terephthalate metal-organic framework (MIL-101) chelated to N,N-dimethylaminopyridine and related aminopyridines. *ACS Applied Materials & Interfaces* 5(4):1269–1278.

Wyszogrodzka, G., B. Marszałek, B. Gil, and P. Dorożyński. 2016. Metal-organic frameworks: Mechanisms of antibacterial action and potential applications. *Drug Discovery Today* 21(6):1009–1018.

Yamashita, M., M. Suzuki, H. Hirai, and H. Kajigaya. 2001. Iontophoretic delivery of calcium for experimental hydrofluoric acid burns. *Critical Care Medicine* 29(8):1575–1578.

Yang, C.-X. and X.-P. Yan. 2011. Metal–organic framework MIL-101(Cr) for high-performance liquid chromatographic separation of substituted aromatics. *Analytical Chemistry* 83(18):7144–7150.

Zhao, T., F. Jeremias, I. Boldog, B. Nguyen, S. K. Henninger, and C. Janiak. 2015. High-yield, fluoride-free and large-scale synthesis of MIL-101(Cr). *Dalton Transactions* 44(38):16791–16801.

Summary

Mesoporous metal–organic frameworks (MOFs) have been a "hot" topic of research in the last 10 years. This book deals with their physical and chemical properties and the capabilities of mesoporous MOFs for adsorption/desorption in solution. Thousands of organic and inorganic compounds are used as feedstock in industrial synthesis and as major components of liquid fossil fuels, dyes for textile coloring, solvents, agricultural pesticides, and pharmaceutical drugs. Adsorption of these compounds in the aqueous and nonaqueous phase is critical in separations, detoxification of waste streams, chromatography, catalysis, administration of medicinal drugs, and beyond.

This book has been written to cover and discuss published experimental research on adsorption and desorption of chemical compounds of various classes on mesoporous MOFs in solution. The topics of primary interest were how the structure of the MOF translates to its adsorption capacity, the selectivity of adsorption, and the regeneration capability of "spent" rather expensive MOF sorbents. Emphasis is placed on molecular mechanisms of adsorption and sorbent regeneration to provide a guide and stimulate future academic research. The capabilities and limitations of mesoporous MOFs as sorbents have been analyzed, along with their potential application areas for chemical industry, petroleum engineering, and environmental and biomedical applications.

We have attempted to conduct a comprehensive review of pertinent literature, covering all relevant applications and common classes of adsorbates of major interest and the contribution from researchers in this field. This book covers more than 300 original publications, mostly research articles in peer-reviewed journals. This book can serve as a reference handbook for readers in various subfields of chemistry, chemical engineering, materials science, biomedical science, and environmental research who work with advanced contemporary porous functional materials. However, this book, as probably any other book, has been written with the limitation of available resources. Our sincere apologies to those whose papers turned out to not be included in this book. Progress in this research field occurs very quickly, and the whole literature on this topic cannot be covered. We sincerely hope that this book will be useful to the scientific and engineering community in academia and industry.

List of Abbreviations

2-ATA	2-Aminoterephthalic acid
1,2DMI	1,2-Dimethylindole
4-MAP	4-Methylaminopyridine
4-MDBT	4-Methyldibenzothiophene
2MI	2-Methylindole
3-MT	3-Methylthiophene
4,6-DMDBT	4,6-Dimethyldibenzothiophene
AAS	Atomic absorption spectroscopy
AC	Activated carbon
acac	Acetylacetonate
AMSA	Aminomethanesulfonic acid
ASA	p-Arsanilic acid
ATR-FTIR	Attenuated reflectance Fourier transform infrared spectroscopy
AZT	Azidothymidine
AZT-TP	Azidothymidine triphosphate
BASF	Badische Anilin und Soda Fabrik
BDC	Benzenedicarboxylic acid
BE	Binding energy
BET	Brunauer–Emmett–Teller method
BIO-MOF	Biocompatible metal-organic framework
BJH	Barrett–Joyner–Halenda method
BPA	Bisphenol A
BSA	Bovine serum albumin
BT	Benzothiophene
BTC	Benzenetricarboxylic acid
CBZ	Carbazole
CDBs	Cis–diol containing biomolecules
CDV	Cidofovir
CR	Congo Red dye
CTAB	Cetyltrimethylammonium bromide
CUS	Coordinatively unsaturated site
DAAP	Dialkylaminopyridine
DBT	Dibenzothiophene
DDS	Drug delivery system
DEG	Diethylene glycol
DETA	Diethylenetriamine
DFT	Density functional theory
DI	Deionized (water)
DLS	Dynamic light scattering
DMA	N,N-dimethylacetamide

DMEM	Dulbecco's modified Eagle's medium
DMEN	N,N-dimethylethylenediamine
DMF	Dimethylformamide
DOX	Doxorubicin
DRS	Diffuse reflectance spectra
DSS	Disuccinimidyl suberate
DTG	Derivative thermogravimetric analysis
ED	Ethylenediamine
EDX	Energy-dispersive x-ray spectroscopy
EG	Ethylene glycol
EN	Ethylenediamine
EPA	U.S. Environmental Protection Agency
EXAFS	Extended x-ray absorption fine structure spectroscopy
FC	Fuel cell
FESEM	Field emission scanning electron microscopy
FTIR	Fourier transform infrared spectroscopy
GFP	Green fluorescent protein
GI-WAXS	Grazing incidence wide-angle x-ray scattering
GO	Graphene oxide
H3BTC	1,3,5-Benzenetricarboxylic acid
H3TATB	Triazine-1,3,5-tribenzoic acid
HAS	Human serum albumin
HDN	Hydrodenitrogenation
HDS	Hydrodesulfurization
HIV	Human immunodeficiency virus
HKUST-1	MOF by the Hong Kong University of Science and Technology
HPLC	High-performance liquid chromatography
HSAB	Hard and soft acids and bases concept
ICP	Inductively coupled plasma
IL	Ionic liquid
IND	Indole
IRMOF	Isoreticular metal–organic framework
IRRAS	Infrared reflection absorption spectroscopy
IUPAC	International Union of Pure and Applied Chemistry
KRICT	Korea Research Institute of Chemical Technology
LD50	Lethal dose, 50%
LMCT	Ligand-to-metal charge transfer
LOD	Limit of detection
MAS NMR	Magic-angle spinning nuclear magnetic resonance spectroscopy
MB	Methylene blue dye
MG	Malachite green dye
MIL	Materials of Institut Lavoisier
MO	Methyl Orange dye
MOF	Metal–organic framework
MOP	Metal-organic polyhedron

MP-11	Microperoxidase-11
NAP	Naphthalene
NB	Nitrobenzene
NMCBZ	N-methyl carbazole
NMOF	Nanocrystalline metal–organic framework
NMR	Nuclear magnetic resonance
NP	Nanoparticle
NSAID	Nonsteroidal antiinflammatory drug
ODS	Oxidative desulfurization
ON	Octane number
OTC	Over the counter
PAHs	Polyaromatic hydrocarbons
PAM	Preassembled modification (method of synthesis of MOFs)
PBS	Phosphate-buffered saline
PCDDs	Polychlorinated dibenzodioxins
PETE	Polyethylene terephthalate
PFOA	Perfluorooctanoic acid
PM	Particulate matter
PNP	p-Nitrophenol
POM	Polyoxometalate
ppm	Parts per million
ppmw	Parts per million by weight
PSM	Postsynthetic modification
PVP	Polyvinylpyrrolidone
PWA	Phosphotungstic acid
PXRD	Powder x-ray diffraction
PYRID	Pyridine
PYRR	Pyrrole
QUIN	Quinoline
RhB	Rhodamine B dye
RH	Relative humidity
ROX	Roxarsone
RT	Room temperature
SAXS	Small-angle x-ray scattering
SBF	Simulated bodily fluid
SBU	Secondary building unit
SEM	Scanning electron microscopy
SGF	Simulated gastric fluid
SIF	Simulated intestinal fluid
SRGO	Straight run gas oil
STM	Scanning tunneling microscopy
SURMOF	Surface-anchored metal–organic framework
TD-DFT	Time-dependent density functional theory
TEG	Triethylene glycol
TEM	Transmission electron microscopy

TGA	Thermogravimetric analysis
Th	Thiophene
THF	Tetrahydrofuran
TIPB	1,3,5-Triisopropylbenzene
TMAOH	Tetramethylammonium hydroxide
TPT	Topotecan
Tris	Tris(hydroxymethyl)aminomethane
TUD	Technische Universität Dresden
UiO-66	MOF by the University of Oslo
ULS	Ultralow sulfur
ULSD	Ultralow sulfur diesel
UMCM-1	University of Michigan Crystalline Material-1 MOF
UVA	Ultraviolet light with $\lambda = 315{-}400$ nm
UV-Vis DRS	UV-Visible diffuse reflectance spectroscopy
XANES	X-ray absorption near-edge structure spectroscopy
XAS	X-ray absorption spectroscopy
XO	Xylenol Orange dye
XPS	X-ray photoelectron spectroscopy
XRD	X-ray diffraction

Index

Printed and bound by CPI Group (UK) Ltd, Croydon, CR0 4YY

01/11/2024

01782619-0004